Vehicular Networks

Vehicular Networks

Models and Algorithms

Edited by
André-Luc Beylot
Houda Labiod

Series Editor
Guy Pujolle

iSTE

WILEY

First published 2013 in Great Britain and the United States by ISTE Ltd and John Wiley & Sons, Inc.

ISTE Ltd
27-37 St George's Road
London SW19 4EU
UK

www.iste.co.uk

John Wiley & Sons, Inc.
111 River Street
Hoboken, NJ 07030
USA

www.wiley.com

Library of Congress Control Number: 2013936314

British Library Cataloguing-in-Publication Data
A CIP record for this book is available from the British Library
ISBN: 978-1-84821-489-7

MIX
Paper from responsible sources
FSC® C013604
FSC www.fsc.org

Printed and bound in Great Britain by CPI Group (UK) Ltd., Croydon, Surrey CR0 4YY

Table of Contents

Introduction

Due to the technical improvements implemented by car manufacturers, we have recently witnessed a significant decrease in road traffic accidents in developed countries. However, there is still considerable scope for improvement in the field of road safety. The advancement made in wireless communications provides numerous possibilities for offering drivers a large panoply of interesting services in the field of intelligent transport systems (ITS). The proposed solutions include the possibility to enable communication directly between vehicles or through a telecommunication infrastructure. The first solutions are thus related to infrastructureless communications and *ad hoc* networks; so we will discuss vehicular *ad hoc* networks (VANETs); in contrast, the second set of solutions comprises more conventional communications that can use infrastructures (general packet radio service (GPRS), universal mobile telecommunications system (UMTS), long-term evolution (LTE), etc.). Hybrid solutions could be involved in order to make the best use of available resources.

Therefore, from a network point of view, we see that new, specific problems are emerging. These problems are related not only to the particular applications implemented but also to the heterogeneous aspect of the types of networks used. For example, in the given context, we cannot simply apply the proposed solutions to *ad hoc* networks (such as mobile *ad hoc* network (MANET)). The tackled themes can be found at the crossroads of several research communities: the telecommunications research community and the research community of transport systems.

In this book, we discuss several interesting and relevant research topics related to vehicular networks, such as congestion control, routing, clustering, interconnection between vehicular networks and LTE/LTE advanced networks, signal traffic control, simulation tools and mobility trace generation.

The main objective of this book is to present the contributions brought by each research community in their respective fields. Finally, we have chosen a descriptive approach to draw up exhaustive reports, to globally present the individual author contributions, to illustrate clearly their advantages and limitations, and to pave the way for future research. Readers wishing to broaden their knowledge of the technical concepts will find at the end of each chapter a set of references and the recent publications of various authors.

Considering the diversity of the fields discussed in various chapters, this book is structured into seven chapters.

Following the Introduction written by Houda Labiod and André-Luc Beylot, Chapter 1 written by Razvan Stanica, Emmanuel Chaput and André-Luc Beylot presents a state of the art of the congestion control protocols in VANET networks. This problem is very critical. A tendency toward decentralized congestion control is emerging at the level of academic research, as well as at the level of standardization, more particularly within the European Telecommunications Standards Institute (ETSI) where several technical specifications have been published on this subject. Five approaches are discussed in this chapter: the first approach based on the adaptation of the sending frequency of beacons, the second approach based on the increase in data transmission flow rate (due to the use of complex modulations), the third approach based on the transmission power control in order to increase the channel capacity, the fourth approach based on the reduction of the contention window size and, finally, the fifth approach based on carrier sensing. A performance assessment of several adaptive mechanisms involved is presented by comparing them to the IEEE 802.11p standard mechanism.

Chapter 2, written by Xunxing Diao, Jian-Jin Li, Kun-Mean Mou and Haiying Zhou, focuses on the geographical routing techniques in a pure VANET. The routing is, of course, a basic, indispensable feature that must be supported by every *ad hoc* network, including VANETs. The routing in vehicular networks – which is different from classic IP routing and from MANET routing – is, in particular, a challenging problem due to the high mobility of vehicles on the one hand and the frailty of wireless connections on the other hand, and due to the strong constraints of the applications as well. The chapter presents a summary of various ITS projects related to intervehicular communications. Wireless technologies, which are indispensable in the design of all routing techniques, are made available, developed and experimented by these ITS projects, and described in detail before addressing the key problem, that is geographical routing dedicated to VANET. In the conclusion of this chapter, the authors sketch a list of open questions such as security, location management, transport layer contextual techniques and, finally, the support of the Quality-of-Service in order to increase the reliability and efficiency of the applications.

Chapter 3, written by Véronique Vèque, Florent Kaisser, Colette Johnen and Anthony Busson, analyzes the forming of clusters in vehicular networks. The authors start out from the assumption that the VANETs by themselves cannot implement all the applications correctly, primarily because of their intermittent connection. They can only function in conjunction with an infrastructure. However, if we observe road traffic, we notice that natural groups of vehicles are formed and the main objective then becomes to take advantage of these geographical characteristics in order to form clusters. The aim of clustering is to facilitate the organization of communications and minimize their cost. The authors then propose a hierarchical protocol called a "convoy", which allows the construction of stable clusters as well as providing scalability.

Chapter 4, written by Guillaume Rémy, Sidi-Mohammed Senouci, François Jan and Yvon Gourhant, sheds more light on the previous chapter by focusing on the complementarity between infrastructureless vehicular networks and LTE networks. The idea is thus to fill in the gaps of the infrastructure-based network coverage by using intervehicle communications. The solution is called LTE for vehicle-to-X communications (LTE4V2X) and has several characteristics. A first protocol allows us to collect information and organize the network in a centralized manner. Depending on the total or partial coverage by the LTE network, several scenarios are considered. A second protocol deals with the dissemination of data toward the vehicles, uniquely either in LTE or in multihop networks. Giving specific examples, the authors show that their solution is powerful and it allows us to fix, quite effectively, the gaps in coverage due to the presence of tunnels, for example.

Chapter 5, written by Ghayet El Mouna Zhioua, Houda Labiod, Nabil Tabbane and Sami Tabbane, discusses the integration of VANET networks into fourth-generation mobile networks. The association between a mobile network and a VANET network aims to improve the coverage of the mobile network and the Quality-of-Service, while having the possibility to resort to alternative traffic routes in case there are any problems on the usual connections. In the first stage, the authors give an overview of the state of the art of clustering algorithms proposed in the relevant literature. The gateway selection problem for the vehicle-to-infrastructure (V2I) connection is discussed in the case of traffic transport from the VANET network toward the infrastructure. The authors study the proposed algorithms in a clustered and non-clustered VANET architecture. Then, the authors look into the problem of gateway selection from the VANET network toward the LTE advanced network.

Chapter 6, written by Jérôme Härri, Sandesh Uppoor and Marco Fiore, deals with the simulation of vehicular networks. The authors present an exhaustive report on the simulation tools used, including microscopic, macroscopic and mesoscopic traffic simulators, as well as on the interactions between these different simulators. The chapter details the trace generation/mobility models used by the network simulators aimed for the assessment of different vehicular networks' mechanisms; it also provides the reader with the basic elements for successfully carrying out simulations for these type of networks.

Chapter 7 describes the signal traffic control systems. The authors provide a classification of the different existing systems and a fine comparison between them. A special emphasis is placed on the dynamic systems whose objective is to reduce traffic jams and improve traffic flow. A new original approach via vehicle-to-vehicle communications is presented. The proposed control system adjusts the duration of traffic lights by using the density information provided by the dissemination protocols, which, in turn, use geographic and directional clustering.

Besides presenting several relevant and very interesting areas of research, we hope that this book will contribute to bring a realistic global view of the evolution of VANETs. As all the authors in this book have already pointed out, there still remain numerous research topics to be explored.

We warmly thank the authors for their very relevant contributions and the quality of their work, as well as the proofreaders who had the difficult task of helping us deliver a final version of this book.

Houda LABIOD and André-Luc BEYLOT
April 2013

Chapter 1

Congestion Control for Safety Vehicular *Ad Hoc* Networks

1.1. Introduction

In the highly dynamic vehicular environment, congestion control is essential, especially with regard to safety messages. Although a dedicated spectrum has been allocated for vehicular communications, the European 30 MHz Intelligent Transportation System (ITS) band (with a possible extension to 50 MHz) or the US 75 MHz Direct Short Range Communication (DSRC) band still represent a scarce resource and need efficient mechanisms in order to be optimally used under high vehicular density. In both Europe and the US, the allocated spectrum has been divided into 10 MHz channels. From these channels, one is known as the control channel (CCH) and it is used solely by road safety applications. The rest of the channels, called service channels (SCH), can be used by both safety and non-safety applications.

Chapter written by Razvan STANICA, Emmanuel CHAPUT and André-Luc BEYLOT.

The number of proposed vehicular safety applications that could use direct vehicle to vehicle (V2V) communication is impressive [PAP 09]. However, at a close inspection, it can be noted that all these applications practically use the same information, coming from onboard sensors of neighboring vehicles: speed, acceleration, steering angle and location.

Considering this, the standardization bodies decided to add a supplementary layer between the applications and the transport protocol. The role of this layer, called *message sublayer* in the IEEE Wireless Access in Vehicular Environments (WAVE) architecture and *facilities layer* in the ETSI ITS terminology, is to keep an accurate image of the surrounding environment inside every vehicle and to provide applications with the desired information.

The facilities layer only needs two types of messages in order to achieve these objectives, called (in the ETSI ITS architecture) cooperative awareness message (CAM) and decentralized environmental notification message (DENM). CAMs are regular beacons, transmitted by every vehicle with a predetermined frequency, and containing details about the vehicle that might be relevant to its neighbors from a safety point of view. In addition, if a vehicle detects a potential hazard (e.g. a sudden brake) and considers that this information needs to be quickly disseminated to the other traffic participants, it transmits a DENM.

However, regardless of the scenario and message type, these safety messages are always transmitted in broadcast mode at the medium access control (MAC) layer. Even in the case when the transmitted information targets a certain geographical area (e.g. an electronic brake alarm is only of interest to vehicles traveling in the same direction as the transmitter and situated behind it), the message is still broadcast and the filtering happens at the facilities layer, as described by the ETSI framework [EUR 10].

The broadcast nature of the CCH in vehicular *ad hoc* networks (VANET) is an essential property that distinguishes it from other IEEE 802.11-based networks. As a matter of fact, the numerous studies on the distributed coordination function (DCF) implementing MAC mechanisms in IEEE 802.11 usually focus on unicast traffic, and broadcast messages are only considered for control purposes. Oliveira et al. [OLI 09] quantify the influence of broadcast traffic on the performance of IEEE 802.11 networks, and they find out that the effect of broadcast messages becomes significant when the proportion of broadcast traffic is higher than 50%. In this scenario, the behavior of the network largely deviates from what is predicted by classic DCF models. However, the authors consider this situation *quite unreal* and they do not investigate the issue further.

Another important characteristic of safety messages comes from the limited lifetime of CAMs. As these beacons are produced periodically by the facilities layer, there is a certain probability that they can expire before the MAC layer has the opportunity to transmit them. When a CAM is waiting for the IEEE 802.11 back-off timer to expire, and the next beacon also arrives in the transmission queue, the first message has to be dropped, as its transmission would only disseminate outdated information to its neighbors. This property, rarely taken into consideration in VANET studies, has a significant effect on the optimal value of different MAC layer parameters.

The IEEE 802.11p amendment [THE 10] is the preferred MAC technology in both the IEEE WAVE and the ETSI ITS architectures. IEEE 802.11p radios can communicate at a distance of 1 km. In a simple scenario, with a two-lane road in both directions and an average inter-vehicular distance of 50 m (a medium density highway), the number of one-hop neighbors reaches 160 vehicles. This is clearly a more challenging environment than the classic Wireless Local Area Network (WLAN), with a central access point and no more than 10–20 nodes. The MAC layer protocol, therefore, needs solutions for this congested environment to achieve scalability.

Congestion control mechanisms received a lot of attention from the VANET research community and the most relevant studies in this area are summarized later. The standardization bodies also recognized the importance of a decentralized congestion control framework for V2V safety communications, and ETSI published a series of technical specifications in this area in July 2011 [EUR 11]. In the US, the Society of Automotive Engineers (SAE) is also developing a standard with similar objectives, SAE J2945.1, currently in a draft phase. SAE J2945.1 is expected to be integrated into the WAVE architecture as a complement for the different IEEE standards.

In this chapter, five different approaches for MAC layer congestion control are discussed. In section 1.2, beaconing frequency adaptation is presented that reduces the number of transmitted safety messages in a dense network, speculating the relationship between high density and reduced speed in vehicular traffic. In section 1.3, increased data rates can be achieved by using more complex modulations and result in a lower occupancy of the CCH. Other proposals form the object of section 1.4, which are based on the fact that transmission power control has an important impact on the number of hidden nodes, and can increase the spatial reuse and hence the channel capacity, in a congested network. In section 1.5, the fourth element, the minimum contention window (CWmin), is analyzed, a parameter with a major importance for collision probability in an IEEE 802.11 network. Finally, the role of the physical carrier sense in congestion control is highlighted in section 1.6.

1.2. Beaconing frequency

The most obvious solution for controlling the channel load in a congested environment is to reduce the number of transmitted messages. This can be achieved in a straightforward manner in vehicular networks by adapting the frequency of the safety beaconing. However, such an adaptive mechanism should be designed carefully because sending less messages can easily have the effect of damaging the performance of safety applications instead of improving it.

In this context, Fukui *et al.* [FUK 02] proposed transmitting a CAM every time the vehicle travels a certain distance instead of using a regular time interval. According to a fundamental relationship from traffic theory, the mean speed decreases when the vehicular density increases, thus the consequence of this approach would be that nodes would reduce the beaconing frequency in a dense network where they would travel at low speeds. However, a basic example for which this solution fails is that of a vehicle waiting to make a left turn in normal traffic. Because the vehicle would need to stop, the adaptive mechanism would practically turn off the beaconing transmission, making an application like the left turn assistant practically unusable. Therefore, as stationary vehicles or low speeds are not always the consequences of high vehicular densities, such an approach cannot be efficiently used in a real scenario.

As a part of the California PATH program, Rezaei *et al.* [REZ 07] take a more complex approach, where vehicles run an estimator to calculate the position of each one-hop neighbor based on the already received messages. The same estimator is used by the node to predict its own position, as it would be calculated by its neighbors. When the difference between the prediction and the actual location becomes larger than a predefined threshold, the node transmits a safety beacon. The problem with this solution is that it is efficient in the predictable free-flow traffic, but not in a congested scenario where the acceleration is highly variable. Moreover, this self-estimator approach does not take into account that the error at some of the neighbors might be considerably different because some of the transmitted beacons could be lost. To solve this problem, Huang *et al.* [HUA 10] further develop this idea using the packet error ratio (PER) measured by a node to predict the losses encountered by its neighbors. Still, measuring a PER in a vehicular network without being able to detect collisions or use feedback from the receivers is not a straightforward task.

Seo *et al.* [SEO 10] make an analogy between the safety beaconing and the coupon collector problem. The mechanism they design relies upon nodes piggybacking acknowledgments (ACKs) for the received beacons in their own safety message. Every received ACK would further delay the transmission of the next CAM, reducing the beaconing frequency. However, the introduced overhead would be significant, especially in a dense network (a 4 byte ACK for 50 one-hop neighbors would result in 200 extra bytes for every safety message). It is also unclear if this approach would be compatible with a security framework based on changing pseudonyms, like the approach currently proposed by the ETSI ITS architecture [PAP 08], because the ACK would need to include the identifier of the sender and most probably a sequence number for the acknowledged message.

Adaptive Traffic Beacon (ATB) is a solution/mechanism/approach proposed by Sommer *et al.* [SOM 11], where the beaconing frequency is calculated based on two metrics: the channel quality and the message utility. The idea is to transmit only the most important messages in a congested network, reducing the offered load. Nevertheless, the channel quality is very sensitive to the number of collisions, which implies that the nodes are somehow supposed to detect such events, clearly a difficult task in a broadcast environment [STA 12]. Moreover, while different utility factors could help differentiate between CAMs and DENs, safety beacons would be difficult to prioritise, as they belong to the same message class. Finally, ATB increases the beaconing period to a mean of 3.6 s, clearly a value that does not comply with the delay requirements of most safety applications, which vary between 100 ms and 500 ms [PAP 09].

For more details on adaptive beaconing solutions, the reader is referred to the very comprehensive review paper by Schmidt *et al.* [SCH 10]. To conclude, while reducing the beaconing frequency is a powerful tool in congestion control, the consequences of this adjustment on every safety application should be taken into account. However, road safety applications will most likely not be standardized, and addressing the constraints imposed by proprietary solutions is a difficult task.

1.3. Data rate

The standards from the IEEE 802.11 family provide multi-rate capability at the physical layer, but without specifying a particular approach for data rate adaptation. In wireless communications, a more complex modulation results in a higher data rate, but it also requires a higher signal-to-noise ratio (SNR) at the receiver in order to be correctly decoded. In the continuous fight for increased bandwidth, the search for an efficient data rate control solution in the very lucrative WLAN industry stimulated the research in this area, and two main classes of mechanisms have been designed.

The solutions in the first class are based on their choice for a certain modulation and coding rate on the success or failure of previously sent messages. For example, the Robust Rate

Adaptation Algorithm (RRAA), proposed by Wong *et al.* [WON 06], calculates the frame loss ratio in a short time window and compares this value with two predefined thresholds. Too many losses determine a reduction in data rate, while a high percentage of successful transmissions results in the choice of a more complex modulation. The second type of mechanisms are based on feedback from the receiver regarding signal quality. A representative example in this class is receiver-based auto rate (RBAR), described by Holland *et al.* [HOL 01]. RBAR relies upon the idea of receivers measuring the channel quality by analyzing the Request To Send (RTS) message and calculating the highest achievable data rate based on the channel conditions. This information reaches the transmitter through the Clear To Send (CTS) message and the best modulation is set for the data frame.

The applicability of mechanisms from the two classes discussed above in a unicast vehicular network is studied experimentally by Camp and Knightly [CAM 08]. They show that, because of the highly variable vehicular channel, decisions based on historical data are not accurate in this environment, while the SNR-based mechanisms need to be trained in the target geographical region in order to cope with the short coherence time (around 300 μs when other vehicles are also present on the road).

In broadcast safety communications, solutions using feedback from the receivers are clearly unsuitable, therefore the data rate adaptation mechanisms proposed for vehicular safety messages follow the classic path of algorithms based on historical data. Mertens *et al.* [MER 08] use RRAA in their simulation study, showing a significant improvement in performance when compared with regular IEEE 802.11p. Nevertheless, they do not address the problem of computing the frame loss ratio in a VANET. A more innovative approach is proposed by Ruffini and Reumerman [RUF 05], building on the correctly received CAMs to create a map of the average path loss at different receivers and use this map to estimate the highest data rate that could be successfully used.

However, the data rate adaptation problem is not exactly equivalent in WLAN and in safety vehicular networks. In the first case, the goal is to maximize throughput by choosing the corresponding modulation. While the problem is, of course, difficult to solve, the existence of a solution cannot be questioned. In a VANET, the goal, as described in the different congestion control architectures, is to reduce the transmission time of a message when the vehicular density increases to give more stations the chance to access the channel during a beacon period. The choice of the modulation is not dictated in this case by the quality of the channel, but by the number of one-hop neighbors, and there is currently no proof that the assignment of a data rate based solely on the local node density could increase the beaconing reception ratio. Moreover, an experimental study led by General Motors R&D and presented by Bai *et al.* [BAI 10] argues that using Quadrature Phase-Shift Keying (QPSK) and a data rate of 6 Mb/s) is the only reasonable choice for V2V communications. In their tests, only two communicating vehicles have been used, ignoring therefore the impact of message collision or interference. Even in these idealistic conditions, any modulation resulting in a higher data rate drastically reduces the reception probability, even at small distances from the transmitter (less than 50% received beacons at 50 m using 18 Mb/s). Furthermore, even the more robust 3 Mb/s Binary Phase-Shift Keying (BPSK) modulation shows lower performance, because, in this case, the transmission time is larger than the coherence time (found to be around 300 μs, just like in [CAM 08]).

Considering these results, data rate adaptation mechanisms need to be better evaluated, especially using real hardware during field tests, before a decision relative to their usefulness in VANET congestion control can be taken.

1.4. Transmission power

Transmission power control is one of the most studied topics in the area of VANET congestion control. However, most of the proposed mechanisms are just variants of solutions previously proposed in a mobile *ad hoc* network (MANET) context, where the objective of adjusting the transmission power is to minimize energy consumption while keeping a connected network. For example, Chigan and Li [CHI 07] use a directional antenna approach originally designed for topology control in MANETs to obtain the minimal power needed to transmit messages only to the closest vehicle on each direction. Similarly, Yoon and Kim [YOO 11] adapt transmission power with the objective of keeping a constant number of one-hop neighbors.

Nevertheless, these solutions are not appropriate for a safety VANET, where messages need to cover a minimal distance, not a certain number of neighbors. With these requirements in mind, Guan *et al.* [GUA 07] define a *target range* for safety messages. When a node receives a message, it calculates the distance from the sender and verifies if it is positioned inside the target range. Vehicles receiving a beacon despite being outside the target range include the identifier of the transmitter in a special feedback field in their own beacon. Using the information in this field, a station can calculate how many nodes outside the target range were reached by its transmission and the goal of the power control mechanism is to keep this number between certain limits.

Another proposal using special feedback piggybacked in the CAMs is the distributed fair power adjustment for vehicular environments (D-FPAV) strategy described by Torrent-Moreno *et al.* [TOR 09]. D-FPAV defines a maximum beaconing load (MBL) that can be accommodated by the CCH while still having spare bandwidth in the eventuality of a special notification. A distributed algorithm ensures an optimal power level assignment, where vehicles use the maximal possible power that still respects the MBL constraint. However, this optimality is achieved only when the power levels used by all the two-hop neighbors are known.

Because the overhead introduced by D-FPAV is significant, especially under high node density when saving bandwidth is the most important, Mittag el al. [MIT 08] designed segment-based power adjustment for vehicular environments (SPAV). SPAV on the one hand does not achieve an optimal assignment like D-FPAV, but on the other hand it does not require full knowledge about the power levels used by different neighbors, but only an estimate of the local density that can be obtained in a much more inexpensive manner.

The local node density (estimated, for example, from the received beacons) is also used in the computation of the transmission power by Rawat *et al.* [RAW 09], but in this case the transmission range is calculated using results from traffic flow theory. Artimy [ART 07] manages to entirely eliminate the overhead for transmission power control, using only data from the onboard speedometer to estimate the local density, again using fundamental relationships from traffic flow theory.

While calculating local density based on the CAMs received from the other vehicles is considered a natural property of the safety beaconing, this task might be complicated by the use of changing pseudonyms. Huang *et al.*

[HUA 10] propose a solution that can cope with the VANET security requirements. In their framework, a node simply measures the channel occupancy from the information provided by the clear channel assignment (CCA) function. If the percentage of time the medium is sensed as busy in the last beaconing period is under a certain threshold $U_{\min}$, the node uses the highest power level, otherwise a linear mapping between the channel occupancy and transmission power is used.

Because of its excellent properties and its feasibility using existing hardware, transmission power control is considered as a central mechanism for congestion control in VANETs and it has been included in the ETSI ITS decentralized congestion control framework [EUR 11].

1.5. Minimum contention window

The minimum contention window ($CW_{\min}$) is one of the most important parameters of the IEEE 802.11 MAC layer. $CW_{\min}$ represents the initial value of the CW, the superior limit of the interval from which the back-off mechanism draws the number of idle slots the station has to wait before attempting a transmission. For unicast communication, the value of CW is doubled every time an expected ACK message is not received within a predefined delay and it is reset to $CW_{\min}$ for every acknowledged reception, leading to the so-called *binary exponential back-off* (BEB) mechanism.

Even before the release of the first version of the IEEE 802.11 standard, Bianchi *et al.* [BIA 96] showed that the optimal value for $CW_{\min}$ depends on the number of contending stations. More exactly, their analysis shows that, in a saturated WLAN, the throughput is maximized when:

$$CW_{\min} \approx n_c\sqrt{2T_t}, \qquad [1.1]$$

where n_c is the number of nodes in the network and T_t is the time needed to transmit the message (acknowledgment included). Building on these results, Cali *et al.* [CAL 00] determined that the protocol's performance peaks when the time the channel is idle due to the back-off mechanism equals the time the channel is occupied by collisions ($T_{\text{idle}} = T_{\text{col}}$).

Despite this well-known property, the IEEE 802.11 standard does not include any mechanism for the adjustment of $CW_{\min}$ when the number of contending stations grows. The main reason for this was that the protocol was designed for WLANs, with a central access point and a limited number of client stations (usually no more than 20) in mind. A second argument came from the use of the RTS/CTS handshake. In this case, collisions are limited to the short RTS and CTS messages, and therefore the time the channel is busy due to collisions is decreased. This implies that, for an optimal functioning, T_{idle} also needs to be reduced, which requires a lower CW. Moreover, with the massive success of multimedia services, and with the introduction of the IEEE 802.11e standard, the minimum CW has been reduced even more, in order to minimize the delay experienced by sensitive video and voice applications. The idea, in this case, was that most users, especially residential users, connect only a reduced number of devices to their access points, and generally use only one or two of them simultaneously. A reduced CW improves the MAC layer performance in this case, while the BEB mechanism is there as a back up for the cases when the number of contending stations increases.

An impressive number of modified back-off mechanisms have been designed in different WLAN scenarios, and Razafindralambo and Valois [RAZ 06] compare the performance of the most significant of these proposals. Most of the solutions considered in this unicast context still require a fixed value for $CW_{\min}$ and only modify the back-off mechanism. For example, Wang *et al.* [WAN 04] argue that when a transmission succeeds after a number of failures it is not correct to reset the CW to its minimal value, because the congestion will continue to exist on the channel. They propose a slower decrease of CW, and only after several acknowledged transmissions in a row. However, Medepalli and Tobagi [MED 06] proved analytically that the impact of $CW_{\min}$ on the throughput of a network is much more significant than the influence of the back-off mechanism.

For more than a decade, all the IEEE 802.11 enhancements related to $CW_{\min}$ adaptation in MANETs belonged to one of the two categories. The methods in the first class (e.g. [KIM 05]) estimate the number of contending stations in the two-hop neighborhood and use equation [1.1] or some variants to calculate the optimal CW. The second type of mechanisms consider the overhead introduced by the local density estimation as prohibitive, and the amount of time the channel is sensed as idle and the number of collisions are measured instead. The CW is adjusted, in this case, in order to keep the equality $T_{\text{idle}} = T_{\text{col}}$ valid: when there are too many collisions on the channel, the back-off time (and with it the idle time) is increased, while when the channel is idle for long time durations, the CW is reduced. A notable example from this second class is IdleSense, proposed by Heusse *et al.* [HEU 05].

Nevertheless, in 2008, Jiang and Walrand [JIA 10] took a completely different approach concerning the back-off mechanism in carrier sense multiple access (CSMA) networks, proposing *optimal CSMA* (oCSMA). The idea behind this new protocol is to adapt the CW of a node as a function of its queue length. In oCSMA, a node begins with an initial value for *contention aggressiveness* (which can be easily translated into a certain $CW_{\min}$) and, when the number of messages in a link queue increases, the transmitter becomes more aggressive in the competition for channel access. Despite having very low complexity and requiring only local information, oCSMA has been proven to achieve throughput optimality under both continuous-time and discrete-time back-off duration [KIM 11], and was implemented using off-the-shelf IEEE 802.11 hardware [NAR 11].

With all these interesting studies coming from related research fields, we might believe that it should be rather straightforward to study and understand the impact of the CW in V2V communications. However, the particularities of the vehicular network translate once again into unique properties that modify the problem entirely. In a VANET, the node density is highly variable and a station can go, within a few minutes, from a vary sparse environment to several hundred contending neighbors. In addition to the fact that the RTS/CTS handshake cannot be implemented and the BEB mechanism is deactivated by the lack of ACK messages, none of the properties that allowed the use of a small CW in IEEE 802.11 WLANs hold in this scenario.

An adaptive mechanism is therefore needed, but a rapid analysis of the compatibility between the solutions described above and the safety VANET shows that the design of this mechanism is not exactly a simple formality. The Bianchi relationship is true for a unicast saturated one-hop WLAN cell, while a safety vehicular network is neither saturated nor fully connected. In addition, collisions remain difficult to detect in V2V communication, therefore IdleSense and other similar approaches cannot be directly transposed in a vehicular environment. Finally, because expired beacons are dropped, the MAC layer always has at most one safety message to transmit [STA 12], and the queue length cannot determine the CW as proposed in oCSMA. Moreover, the goal of all these mechanisms is to maximize throughput, an objective that is not shared by a safety vehicular network.

The few proposals for CW adaptation issued from the VANET research community failed to consider these important differences. Rawat *et al.* [RAW 09] propose a heuristic based on the number of detected collisions, where the CW is increased if the number of collided messages is higher than a predefined threshold. However, the threshold does not depend on the local node density and the technique used for collision detection is not described. The same critique applies to Mertens *et al.* [MER 08], who, in a first phase, estimate the local node density and directly use this result in equation [1.1]. Then, they further refine the value of the CW by increasing $CW_{\min}$ when the percentage of lost beacons becomes higher than a target PER. Balon and Guo [BAL 06] address this issue of measuring the percentage of lost beacons by using the sequence numbers inside the safety messages, which might not be compatible with a privacy framework based on pseudonyms.

In a similar manner, Wang *et al.* [WAN 08] design a heuristic relying upon the channel busy time measured by the CCA function during a predefined time period. In their solution, if the channel busy time increases between two consecutive measures, the CW grows linearly with the observed difference. In the opposite case, CW is reduced, also using a linear relationship. Although the efficiency of this mechanism depends on the initial value of the CW, the authors do not provide any guidelines for the choice of this parameter.

Meanwhile, Jang and Feng [JAN 10] establish a relationship between the number of contending stations and the optimal back-off time in a vehicular network, but their study is focused on unicast communication using RTS and CTS control messages. Finally, Alapati *et al.* [ALA 10] try to maximize throughput using a type of probing mechanism, where the node tests different values for $CW_{\min}$ until an optimum is reached. The problem comes from the fact that this optimum depends on the local density that might vary faster than the convergence speed of the algorithm.

On the basis of these observations, five different mechanisms for CW control in a vehicular environment have been adapted from solutions proposed in the research literature, but not necessarily related to $CW_{\min}$ adjustment. The properties and feasibility of these mechanisms are characterized below, followed by the results of a simulation study using the Java in simulation time/scalable wireless *ad hoc* network simulator (JiST/SWANS) framework [JAV] and the street random waypoint (STRAW) mobility model [CHO 05].

– *Beacon-based neighbor estimation:* Beaconing represents a native method for estimating the number of local neighbors in a VANET. Mertens *et al.* [MER 08] propose to calculate the number of surrounding vehicles by counting the different sources from which at least a beacon has been received in the last T_{update} seconds. However, as discussed above, the number of neighbors from which a beacon was received, $\tilde{n}_c$, determined this way, cannot be directly applied in Bianchi's equation, even though T_t would be very easy to calculate for fixed-size CAMs. In this case, not only does the VANET not correspond to the original assumptions of a full connected saturated network, but also the accuracy of the estimation $\tilde{n}_c$ depends on the beaconing reception ration.

Therefore, instead of using equation [1.1] directly, the first studied mechanism keeps this linear dependency, but uses a more general formula to calculate the CW:

$$CW = \lambda \tilde{n}_c,$$

where λ is a parameter depending on the size of the beacon, and whose optimal value was explored through simulation.

This solution would be relatively simple to implement because the addresses of the neighbors will be stored anyway for routing purposes, therefore a simple counter is needed. Nevertheless, repeatedly changing the pseudonyms of a node for privacy purposes could have a non-negligible impact on the performance of this solution (and also on the routing protocol).

– *Collided packets estimation:* The second mechanism follows the idea of Balon and Guo [BAL 06] and estimates the PER based on a sequence number added to each CAM (if beacons are generated periodically, the same result can be obtained by simply subtracting the number of received messages from the number of one-hop neighbors that can be estimated as described for the previous mechanism). The CW is initially set to a default value ($CW(0) = CW_{\mathrm{def}}$) and, every T_{update} seconds, it is updated using the following algorithm:

$$CW(t) = \begin{cases} min(2 * CW(t-1), CW_{\max}), & if\ \ PER < PER_{\min} \\ \\ max(CW(t-1)/2, CW_{\min}), & if\ \ PER > PER_{\max} \end{cases}$$

The main advantage of this mechanism is that it tries to directly optimize the percentage of delivered beacons. However, the coexistence of a solution based on sequence numbers and a security protocol using changing pseudonyms appears to be extremely difficult. When a vehicle would change its identifiers, it would also need to reset its sequence counter and, therefore, tracking the lost beacons at the receiver level would become an important problem.

– *Idle time counting:* The next studied solution aims at preserving the equality $T_{\mathrm{idle}} = T_{\mathrm{col}}$. To estimate T_{col} in a broadcast environment where collisions cannot be detected, the number of lost beacons, calculated as in the previous mechanism, is used. However, not all the missing beacons are lost because of a collision. A radio propagation problem on the channel or node mobility can produce a similar effect. Because of this, to achieve a better estimation of T_{col}, only the beacons sent by vehicles situated at a distance of less than d_{col} are taken into consideration. Missing messages sent from a geographically close node have a high probability to be lost due to a collision.

As in the previous algorithm, $CW(0) = CW_{\text{def}}$ and a new value is computed every T_{update} seconds:

$$CW(t) = \begin{cases} min(2 * CW(t-1), CW_{\max}), & if\ T_{\text{col}} > \alpha * T_{\text{idle}} \\ \\ max(CW(t-1)/2, CW_{\min}), & if\ T_{\text{idle}} > \alpha * T_{\text{col}} \end{cases},$$

where $\alpha > 1$ is a parameter whose value was explored through simulation.

The simple idea behind this solution is that if there are too many colliding messages, the CW should be increased, while CW should be decremented when the channel is idle for an important amount of time. This proposal presents the same advantages and drawbacks as the previous proposal. A supplementary implementation problem could come from the fact that the station requires the capacity to measure T_{idle}, a feature that is not currently available in all IEEE 802.11 devices.

– *Stop time neighbor estimation:* Although VANETs are built on top of the already existing transportation system, very few V2V communication solutions attempt to profit from ideas investigated in related vehicular research fields, such as, for example, traffic flow theory. The fundamental relationship explored in traffic flow theory describes the dependency between vehicular flow (vehicles/hour/lane), vehicular density (vehicles/km/lane) and speed (km/hour). Local density could, therefore, be estimated using this type of calculation, which does not require the exchange of any message.

The following algorithm is inspired by the approach taken in [ART 07] for transmission power control, where the vehicular density is estimated based on the time the car stops in traffic. Therefore, the vehicle needs to measure the stop time (T_{stop}) in the last T_{update} time window. If $T_{\text{stop}} = 0$, the traffic is in a free-flow state and the CW is set to $CW_{\min}$. If $T_{\text{stop}} = T_{\text{update}}$, the vehicle is considered to be a part of a traffic jam and $CW = CW_{\max}$. For intermediate values, the following formula is used:

$$CW = (T_{\text{stop}}/T_{\text{update}})(CW_{\max} - CW_{\min}) + CW_{\min}$$

The mechanism could be implemented without any additional hardware, as the stop time can already be calculated using data from the speedometer. The problem could lie in the fact that a vehicle might stop for several other reasons than a traffic jam, especially in an urban scenario

– *Speed-based neighbor estimation:* A more accurate estimation of local density based on traffic-flow theory is proposed by Shirani *et al.* [SHI 09], who use vehicle speed and jerk (the derivative of acceleration with respect to time) to adjust the transmission power. Therefore, this final mechanism calculates the local density and the CW as follows:

$$CW = \frac{D_l}{D_{\max}}(CW_{\max} - CW_{\min}) + CW_{\min}$$

where $D_l = |jerk|/speed$, and $D_{\max}$ is the predefined upper threshold.

Although this approach uses more information than the previous approach, it still lacks the ability to handle, without any delay, some situations common to city traffic, which result in a low speed without necessarily implying a high vehicular density (e.g. left-turn and stop sign). Moreover, jerk is not currently measured on a regular basis in vehicles.

The behavior of these five mechanisms has been compared with the original IEEE 802.11p under medium and high vehicular density, with a traffic volume varying between 40 vehicles/km and 80 vehicles/km. In a first phase, simulations were run in order for each of the above algorithms to be optimized with regard to the different parameters that have an impact on their performance. The tested values and those that achieved the best performance for each parameter are shown in Table 1.1. In the case of IEEE 802.11p, a value of 7 has been used for $CW_{\min}$. This is equivalent to the value mentioned in the standard for the access category with the second highest priority [THE 10].

parameter	Tested values	Optimal value
T_{update}	1 s, 5 s, 10 s, 20 s	5 s
λ	0.25, 0.5, 1, 2	0.5
$PER_{\min}/PER_{\max}$	5%/10%, 10%/15%, 15%/20%	5%/10%
α	1, 1.1, 1.2, 1.3, 1.4	1.1
d_{col}	100 m, 150 m, 200 m, 250 m	200 m
$CW_{\max}$	40, 50, 60	50
$D_{\max}$	2, 2.5, 3, 3.5	3

Table 1.1. *Optimal values for the different parameters of the algorithms*

The metric used for this optimization and for the comparison was the beaconing reception probability at a distance of less than 200 m from the source vehicle. This metric is considered to be particularly suitable because it includes in a single value both the probability of collision and the probability of an expired beacon. Moreover, because a beacon that cannot be sent with the required time delay expires, all the received messages respect the imposed time constraints and therefore the analysis of the average delay becomes less important.

The beaconing reception probabilities for the studied solutions can be observed in Figure 1.1. To better understand the behavior of each mechanism, Figure 1.2 shows the average CW as a function of vehicular density.

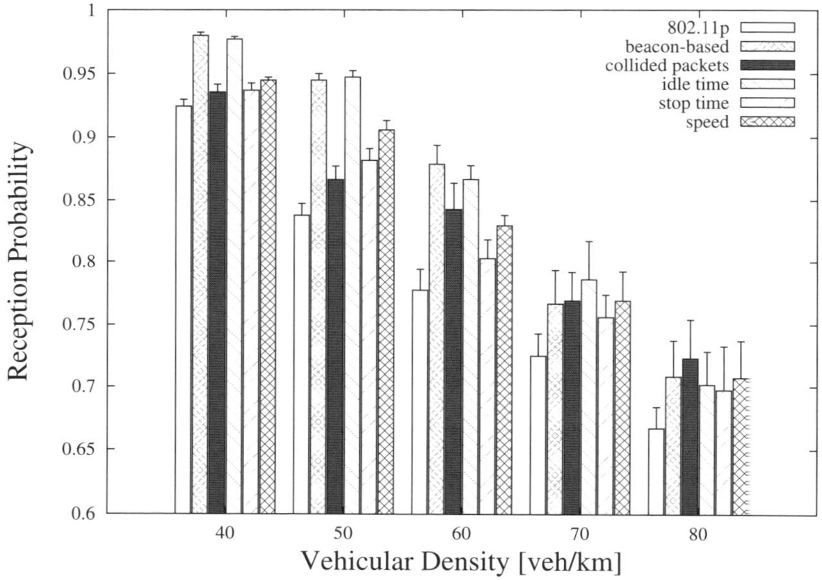

Figure 1.1. *Beaconing reception probability (including expired beacons) using IEEE 802.11p and the five mechanisms described in this chapter for different vehicular densities (the 95% confidence interval is also shown)*

The first thing that can be noted from the data is that all the five solutions show better performance than the basic mechanism, with a difference that can reach more than 10%. Estimating the number of neighbors using the received beacons gives the best results for a vehicular density under 60 cars/km. The idle time approach also gives similar results, showing that the estimation used for T_{col} is quite accurate. However, even though the beacon-based and idle time algorithms show similar results for the CAM reception probability, they achieve this through different means, as it can be seen from the dissimilar average values of the CW. As expected, the beacon-based approach shows a linear increase of CW with the number of vehicles. Therefore, under high density, the nodes back-off for an important amount of time, which leads to an increased number of expired beacons. Moreover, when using the idle time mechanism, the CW converges to an average value of only 36. In this case, the majority of lost messages are due to collisions and the number of expired CAMs is much lower.

An interesting result is obtained when adjusting CW based on the number of lost packets. When the PER is below $PER_{\max}$ (10% in this case), the algorithm almost always uses the default value for CW ($CW_{\text{average}} = 7.5$) and its results are similar with those of the basic IEEE 802.11p. However, when the vehicular density increases and more beacons start colliding, the mechanism starts increasing the CW and its performance drastically improves, showing the best results for a density of 80 vehicles/lane/km. Using a lower value for $PER_{\max}$ increases the efficiency of this solution in low density, it also highly degrades its performance when the number of vehicles becomes more important, indicating that dynamic thresholds would be an interesting approach in this case.

The two solutions inspired from traffic-flow theory also perform better than IEEE 802.11p. However, for the lower values of the vehicular density, their results are not as good as those of the beacon-based or idle sense mechanisms. This is because, in these traffic conditions, vehicles are usually in a free-flow state and they rarely stop or modify their speed in order to increase their CW. Nevertheless, when the number of cars increases, the mobility pattern is also altered and the two algorithms show similar results with the other strategies. It is also important to note that the approach based on cars' speed and jerk always achieves a better reception probability than the approach based on stop time, because it uses a more detailed relationship between car movement and density. Moreover, a very significant property of these two mechanisms is that they can be used together with security solutions based on pseudonyms.

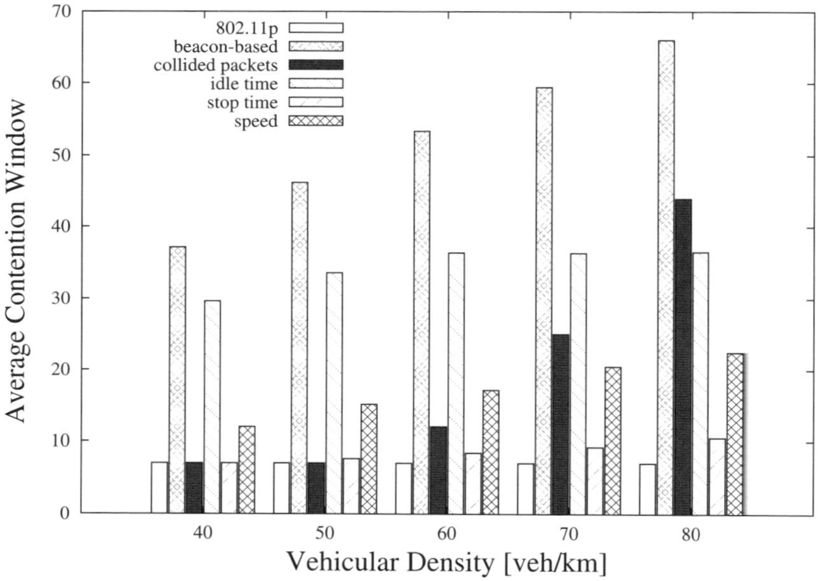

Figure 1.2. *Average contention window (CW) as a function of vehicular density for the analyzed mechanisms*

This comparative study shows the importance that the *CW* has on the MAC layer performance and argues against the use of a small back-off interval in the case of vehicular safety messages, where reliability is an essential parameter and collisions need to be avoided as much as possible.

1.6. Physical carrier sense

The physical carrier sense mechanism is the core of any CSMA-based channel access technique, including the protocols from the IEEE 802.11 family. The concept is well-known and it is used in both wired and wireless networks: before a transmission, a node has to first *sense* the channel to make sure that it is not already occupied by another station.

The carrier sense method described in the IEEE 802.11 standard is based on two functions: Clear Channel Assignment (CCA) and Network Allocation Vector (NAV). NAV is also known as *virtual carrier sense* and is a MAC layer mechanism that uses special control messages – RTS and CTS – in order to reserve the medium for data transmission. CCA is a function of the Physical Layer Convergence Procedure (PLCP) layer and is in charge of physical carrier sensing.

In the case of the Orthogonal Frequency Division Multiplexing (OFDM) physical layer, CCA uses two mechanisms to assess the state of the channel: header detection and energy detection. The PLCP header, shown in Figure 1.3, is always sent using the most robust combination of modulation and coding rate. It contains information on the data rate used for the rest of the message and a LENGTH field indicating the number of bytes to be transmitted. A node capable of decoding the PLCP header calculates the time duration for which the channel will be occupied by this transmission and declares the channel busy for this entire duration, even if the reception of the rest of the message fails. If no PLCP header is detected, the CCA function measures the energy level present on the channel and compares it with a predefined value, called energy detection threshold (ED_t). If the perceived energy level is larger than ED_t, CCA declares the channel busy and denies any MAC layer transmission. To give a numerical example, in the IEEE 802.11p OFDM PHY, the receiver must have the capacity to detect any PLCP header arriving with a power level over −85 dBm and, if the PLCP header is missed, an ED_t of −65 dBm is used.

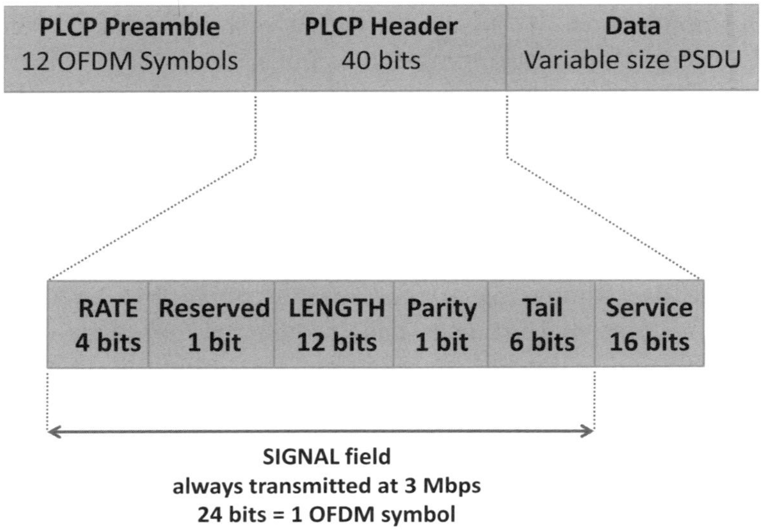

Figure 1.3. *Physical Layer Convergence Procedure (PLCP) header format*

Despite the fact that the physical carrier sensing lays the foundation of an entire category of channel access methods, its impact on the MAC layer performance has received only little attention, at least compared with the recognition received by other parameters, such as transmission power or data rate.

One of the first studies that focused on physical carrier sensing in multi-hop networks was proposed by Zhu *et al.* [ZHU 04], who calculate the optimal carrier sense range for different networks with regular topologies. Their results demonstrate a relationship between the sensing threshold and the signal-to-interference ratio needed to decode the message. Soon afterwards, Yang and Vaidya [YAN 05] pointed out that the value of this optimal carrier sense threshold is higher if the PLCP header is transmitted using the minimum data rate considered. An essential finding came from Kim *et al.* [KIM 08], who discovered that the capacity of a multi-hop wireless network depends only on the ratio between transmission power and carrier sense threshold. In a follow-up of this study, Yang *et al.* [YAN 07] propose a mechanism for topology control through joint transmission power and carrier sense adaptation.

Nevertheless, all these studies have been conducted under the assumption of a pairwise interference model, unrealistic for safety V2V communications. Recently, Fu *et al.* [FU 10] calculated a safe carrier-sensing range that guarantees interference-safe transmissions under the cumulative interference model. The authors also propose a new mechanism, Incremental-Power Carrier-Sensing (IPCS), where the CCA is based on the history of the sensed power level and the medium is declared idle if a sufficient drop in the energy level is detected on the channel. Finally, following an experimental study with an indoor IEEE 802.11 testbed, Brodsky and Morris [BRO 09] conclude that a fixed carrier sense threshold is sufficient in short range networks (under 100 m wide), but it highly degrades MAC layer performance in long range networks, category that clearly includes VANETs.

In a similar case with the other congestion control mechanism, all these ideas are focused on maximizing throughput in wireless networks that do not exhibit the same properties as a safety VANET. The only existing results in the case of a vehicular network come from Schmidt *et al.* [SCH 10], who consider that in the case of safety communications the receivers should be more sensitive, using a lower carrier sense threshold that would allow them to detect transmissions even from vehicles situated far away.

However, this assumption is not supported by any theoretical argument. An increased sensitivity would result in a higher carrier sense range, and therefore in more contending neighbors. The nodes would sense the channel as busy for a longer period and simultaneous transmissions would also be more probable.The remainder of this section is dedicated to a series of simulation results demonstrating the impact of the physical carrier sense threshold on the CAM reception ratio.

This simulation study measures the beaconing reception ratio for three different mean vehicular densities (25 veh/lane/km, 35 veh/lane/km and 45 veh/lane/km) while varying the carrier sense threshold (CS_t) between -95 dBm and -55 dBm. The noise level assumed for these simulations peaks at -98 dBm, and an SNR ratio of at least $3dB$ is necessary for decoding the PLCP header, hence the minimal value of -95 dBm for CS_t. This noise level can be considered relatively high, and the VANET physical channel is indeed very noisy, as it has been confirmed by experimental studies (e.g. [BAI 10]).

In Figure 1.4 the results obtained for a vehicular density of 35 veh/lane/km and three different carrier sense thresholds (-95 dBm, -79 dBm and -71 dBm) are shown. As expected, in all the cases the beaconing reception probability decreases with the distance from the sender. However, comparing what happens when CS_t is modified from -95 dBm to -79 dBm, it can be observed that the reception ratio increases at a distance of less than 200 m from the sender and decreases beyond this distance.

This behavior can be explained by understanding that a collision can be the consequence of a simultaneous transmission (beginning exactly in the same slot) with a station from the carrier sense range, or the result of a concurrent transmission (the messages superpose on at least a slot) with a hidden node. A higher CS_t reduces the number of sensed vehicles, and therefore the probability to consider the medium as busy. This allows more transmission opportunities for every node and reduces the number of simultaneous transmissions. However, on the negative side, because the carrier sense range is reduced, the degree of spatial reuse is increased and therefore concurrent transmission can occur from closer vehicles. This has little effect in the immediate neighborhood due to the capture effect, but produces collisions at higher distances.

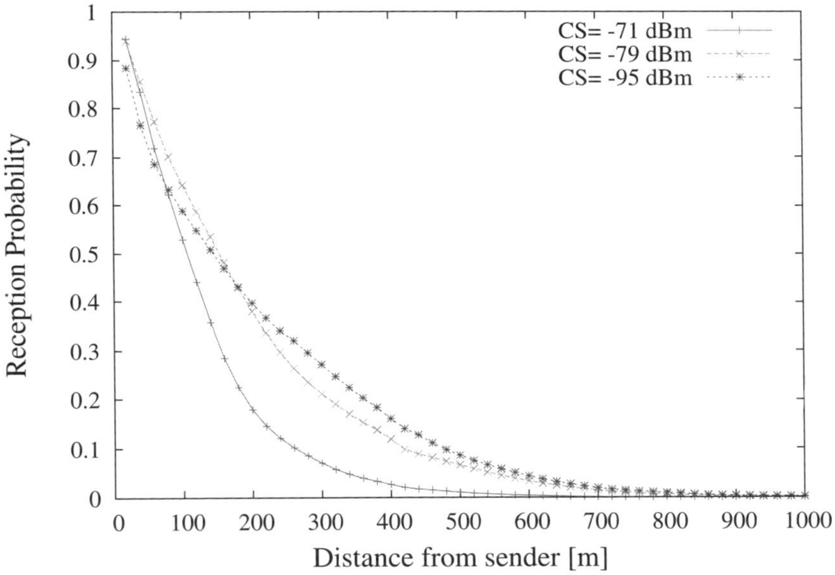

Figure 1.4. *Beaconing reception probability as a function of the distance from the sender for several carrier sense thresholds. The mean vehicular density in this case is 35 veh/lane/km. The 95% confidence interval has about the same size as the symbols and has not been included for visibility purposes*

This phenomenon is exacerbated by further increasing CS_t (from a value of -79 dBm to -71 dBm in Figure 1.4). As the physical carrier sense covers less and less space, the interferer gets closer to the sender and the Signal to

Interference Ratio (SIR) increases, reducing the beaconing reception probability even for closely situated vehicles.

To better understand the influence of the carrier sense threshold on the reception of safety messages, the beaconing reception ratio at 50 m from the transmitter as a function of CS_t for different vehicular densities is shown in Figure 1.5. It can be noted that the number of received beacons slowly increases with CS_t, it reaches an optimal point and then drops quite sharply.

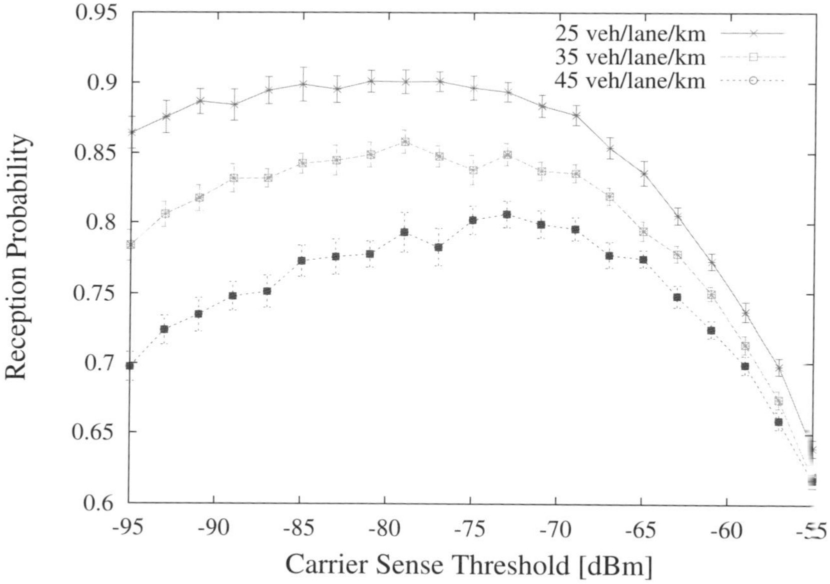

Figure 1.5. *Beaconing reception probability as a function of the carrier sense threshold for different vehicular densities at a distance of 50 m from the sender. The 95% confidence intervals are also shown*

Two other important observations need to be made at this point, challenging the current view of using the minimum receiver sensitivity as a carrier sense threshold [SCH 10] and arguing in favor of a more elaborate solution. First, the difference between the peak value of the reception probability and the one obtained using the lowest possible value for CS_t (-95 dBm in this case) can be significant, reaching almost 10% in the scenario with the highest density. Second, the optimal carrier sense threshold varies with the vehicular density, increasing when the number of neighbors becomes larger. This confirms the ideas formulated earlier in this chapter and shows the necessity of an adaptive mechanism for physical carrier sense control.

1.7. Conclusion

In this chapter, different techniques for congestion control in vehicular *ad hoc* networks have been described. These mechanisms can be divided into two classes: protocol agnostic solutions, which could work with any MAC protocol, and IEEE 802.11 dedicated solutions, designed specifically for the protocols in this family.

The chapter begins with a discussion on generic mechanisms, such as adaptive beaconing frequency, data rate control or transmission power adjustment. The most significant studies in each category are presented and the feasibility of these ideas is commented.

However, the focus of the chapter is on two parameters with a central role in IEEE 802.11, namely the CW of the back-off mechanism and the carrier sense threshold. Adaptive mechanisms for both of these parameters are presented and their performance is compared with the current version of the IEEE 802.11p standard. The improvement brought by choosing the correct value for these parameters is an important argument supporting the design of efficient mechanisms and their integration in the IEEE 802.11p protocol.

1.8. Bibliography

[ALA 10] ALAPATI J., PANDYA B., MERCHANT S., *et al.*, "Back-off and retransmission strategies for throughput enhancement of broadcast transmissions in 802.11p", *Proceedings of the 21st IEEE Intelligent Vehicles Symposium (IV 2010)*, San Diego, pp. 700–705, June 2010.

[ART 07] ARTIMY M., "Local density estimation and dynamic transmission-range assignment in vehicular *ad hoc* networks", *IEEE Transactions on Intelligent Transportation Systems*, vol. 8, no. 3, pp. 400–412, 2007.

[BAI 10] BAI F., STANCIL D., KRISHNAN H., "Toward understanding characteristics of dedicated short range communications (DSRC) from a perspective of vehicular network engineers", *Proceedings of the 16th ACM Annual International Conference on Mobile Computing and Networking (MOBICOM 2010)*, Chicago, pp. 329–340, September 2010.

[BAL 06] BALON N., GUO J., "Increasing broadcast reliability in vehicular ad hoc networks", *Proceedings of the 3rd ACM International Workshop on Vehicular Ad Hoc Networks (VANET 2006)*, Los Angeles, pp. 104–105, 2006.

[BIA 96] BIANCHI G., FRATTA L., OLIVERI M., "Performance evaluation and enhancement of the CSMA/CA MAC protocol for 802.11 wireless LANs", *Proceedings of the 7th IEEE International Symposium on Personal, Indoor and Mobile Radio Communications (PIMRC 1996)*, Taipei, Taiwan, pp. 392–396, October 1996.

[BRO 09] BRODSKY M., MORRIS R., "In defense of wireless carrier sense", *Proceedings of the ACM SIGCOMM Conference on Data Communication (SIGCOMM 2009)*, Barcelona, Spain, pp. 147–158, August 2009.

[CAL 00] CALI F., CONTI M., GREGORI E., "Dynamic tuning of the IEEE 802.11 protocol to achieve a theoretical throughput limit", *IEEE/ACM Transactions on Networking*, vol. 8, no. 6, pp. 785–799, 2000.

[CAM 08] CAMP J., KNIGHTLY E., "Modulation rate adaptation in urban and vehicular environments: cross-layer implementation and experimental evaluation", *Proceedings of the 14th ACM Annual International Conference on Mobile Computing and Networking (MOBICOM 2008)*, San Francisco, pp. 315–326, September 2008.

[CHI 07] CHIGAN C., LI J., "A delay-bounded dynamic interactive power control algorithm for VANETs", *Proceedings of the 42nd IEEE International Conference on Communications (ICC 2007)*, Glasgow, pp. 5849–5855, June 2007.

[CHO 05] CHOFFNES D., BUSTAMANTE E., "An integrated mobility and traffic model for vehicular wireless networks", *Proceedings of the 2nd ACM International Workshop on Vehicular Ad Hoc Networks (VANET 2005)*, Cologne, pp. 69–78, September 2005.

[EUR 10] EUROPEAN TELECOMMUNICATIONS STANDARDS INSTITUTE (ETSI), Intelligent Transport Systems (ITS) – vehicular communications – basic set of applications – part 1: functional requirements, ETSI TS 102 637-1 v1.1.1, September 2010.

[EUR 11] EUROPEAN TELECOMMUNICATIONS STANDARDS INSTITUTE (ETSI), ETSI TS 102 687 v1.1.1 – Intelligent Transport Systems (ITS) – decentralized congestion control mechanisms for intelligent transport systems operating in the 5 GHz range – access layer part, July 2011.

[FU 10] FU L., LIEW S., HUANG J., "Effective carrier sensing in CSMA networks under cumulative interference", *Proceedings of the 29th IEEE Annual International Conference on Computer Communications (INFOCOM 2010)*, San Diego, pp. 1–9, March 2010.

[FUK 02] FUKUI R., KOIKE H., OKADA H., "Dynamic Integrated Transmission Control (DITRAC) over Inter-Vehicle Communications in ITS", *Proceedings of the IEEE 55th Vehicular Technology Conference (VTC Spring 2002)*, Birmingham, pp. 483–487, May 2002.

[GUA 07] GUAN X., SENGUPTA R., KRISHNAN H., *et al.*, "A feedback-based power control algorithm design for VANET", *Proceedings of the 1st IEEE Workshop on Mobile Networks for Vehicular Environments (MOVE 2007)*, Anchorage, pp. 67–72, May 2007.

[HEU 05] HEUSSE M., ROUSSEAU F., GUILLIER R., *et al.*, "Idle sense: an optimal access method for high throughput and fairness in rate diverse wireless LANs", *Proceedings of the ACM SIGCOMM Conference on Data Communication (SIGCOMM 2005)*, Philadelphia, pp. 121–132, August 2005.

[HOL 01] HOLLAND G., VAIDYA N., BAHL P., "A rate-adaptive Mac protocol for multi-hop wireless networks", *Proceedings of the 7th Annual International Conference on Mobile Computing and Networking (MOBICOM 2001)*, Rome, Italy, pp. 236–251, July 2001.

[HUA 10] HUANG C., FALLAH Y., SENGUPTA R., *et al.*, "Adaptive intervehicle communication control for cooperative safety systems", *IEEE Network*, vol. 24, no. 1, pp. 6–13, 2010.

[JAN 10] JANG H.-C., FENG W.-C., "Network status detection-based dynamic adaptation of contention window in IEEE 802.11p", *Proceedings of the IEEE 71st Vehicular Technology Conference (VTC Spring 2010)*, Taipei, Taiwan, pp. 1–5, 2010.

[JAV] JAVA IN SIMULATION TIME FRAMEWORK (JiST), "The Scalable Wireless Ad hoc Network Simulator (SWANS)", http://jist.ece.cornell.edu/.

[JIA 10] JIANG L., WALRAND J., "A distributed CSMA algorithm for throughput and utility maximization in wireless networks", *IEEE/ACM Transactions on Networking*, vol. 18, no. 3, pp. 960–972, 2010.

[KIM 05] KIM M.-S., KWON D.-H., SUH Y.-J., "Adaptive window mechanism for the IEEE 802.11 MAC in wireless *ad hoc* networks", *Proceedings of the 19th International Conference on Information Networking (ICOIN 2005)*, Jeju, South Korea, pp. 31–40, February 2005.

[KIM 08] KIM T.-S., LIM H., HOU J., "Understanding and improving the spatial reuse in multihop wireless networks" *IEEE Transactions on Mobile Computing*, vol. 7, no. 10, pp. 1200–1212, 2008.

[KIM 11] KIM T., NI J., SRIKANT R., *et al.*, "On the achievable throughput of CSMA under imperfect carrier sensing", *Proceedings of the 30th IEEE Annual International Conference on Computer Communications (INFOCOM 2011)*, Shanghai, China, pp. 1674–1682, April 2011.

[MED 06] MEDEPALLI K., TOBAGI F., "Towards performance modeling of IEEE 802.11 based wireless networks: a unified framework and its applications", *Proceedings of the 25th IEEE Annual International Conference on Computer Communications (INFOCOM 2006)*, Barcelona, Spain, pp. 1–12, April 2006.

[MER 08] MERTENS Y., WELLENS M., MAHONEN P., "Simulation-based performance evaluation of enhanced broadcast schemes for IEEE 802.11-based vehicular networks", *Proceedings of the IEEE 67th Vehicular Technology Conference (VTC Spring 2008)*, Singapore, pp. 3042–3046, May 2008.

[MIT 08] MITTAG J., SCHMIDT-EISENLOHR F., KILLAT M., *et al.*, "Analysis and design of effective and low-overhead transmission power control for VANETs", *Proceedings of the 5th ACM International Workshop on Vehicular Inter-Networking (VANET 2008)*, San Francisco, pp. 39–48, September 2008.

[NAR 11] NARDELLI B., LEE J., LEE K., *et al.*, "Experimental evaluation of optimal CSMA", *Proceedings of the 30th IEEE Annual International Conference on Computer Communications (INFOCOM 2011)*, Shanghai, pp. 1188–1196, April 2011.

[OLI 09] OLIVEIRA R., BERNARDO L., PINTO P., "The influence of broadcast traffic on IEEE 802.11 DCF networks", *Computer Communications*, vol. 32, no. 2, pp. 439–452, 2009.

[PAP 08] PAPADIMITRATOS P., BUTTYAN L., HOLCZER T., *et al.*, "Secure vehicular communication systems: design and architecture", *IEEE Communications Magazine*, vol. 46, no. 11, pp. 100–109, 2008.

[PAP 09] PAPADIMITRATOS P., DE LA FORTELLE A., EVENSSEN K., *et al.*, "Vehicular communication systems: enabling technologies, applications, and future outlook on intelligent transportation", *IEEE Communications Magazine*, vol. 47, no. 11, pp. 84–95, November 2009.

[RAW 09] RAWAT D., YAN G., POPESCU D., *et al.*, "Dynamic adaptation of joint transmission power and contention window in VANET", *Proceedings of the IEEE 70th Vehicular Technology Conference (VTC Fall 2009)*, Anchorage, pp. 1–5, September 2009.

[RAZ 06] RAZAFINDRALAMBO T., VALOIS F., "Performance evaluation of backoff algorithms in 802.11 *ad hoc* networks", *Proceedings of the 3rd ACM International Workshop on Performance Evaluation of Wireless Ad Hoc, Sensor and Ubiquitous Networks (PE-WASUN 2006)*, Malaga, Spain, pp. 82–89, October 2006.

[REZ 07] REZAEI S., SENGUPTA R., KRISHNAN H., "Reducing the communication required by DSRC-based vehicle safety systems", *Proceedings of the IEEE 10th International Conference on Intelligent Transportation Systems (ITSC 2007)*, Seattle, pp. 361–366, October 2007.

[RUF 05] RUFFINI M., REUMERMAN H.-J., "Power-rate adaptation in high-mobility distributed *ad hoc* wireless networks", *Proceedings of the IEEE 61st Vehicular Technology Conference (VTC Spring 2005)*, Stockholm, pp. 2299–2303, May 2005.

[SCH 10] SCHMIDT R., LEINMULLER T., SCHOCH E., *et al.*, "Exploration of adaptive beaconing for efficient intervehicle safety communication", *IEEE Network*, vol. 24, no. 1, pp. 14–19, 2010.

[SCH 10] SCHMIDT R., LEINMULLER T., BODDEKER B., *et al.*, "Adapting the wireless carrier sensing for VANETs", *Proceedings of the 7th International Workshop on Intelligent Transportation (WIT 2010)*, Hamburg, Germany, pp. 1–6, March 2010.

[SEO 10] SEO H., YUN S., KIM H., "Solving the coupon collector's problem for the safety beaconing in the IEEE 802.11p WAVE", *Proceedings of the IEEE 72nd Vehicular Technology Conference (VTC Fall 2010)*, Ottawa, pp. 1–6, September 2010.

[SHI 09] SHIRANI R., HENDESSI F., GULLIVER A., "Store-carry-forward message dissemination in vehicular *ad hoc* networks with local density estimation", *Proceedings of the 70th IEEE Vehicular Technology Conference (VTC Fall 2009)*, Anchorage, pp. 1–6, September 2009.

[SOM 11] SOMMER C., TONGUZ O., DRESSLER F., "Traffic information systems: efficient message dissemination via adaptive beaconing", *IEEE Communications Magazine*, vol. 49, no. 5, pp. 173–179, 2011.

[STA 12] STANICA R., CHAPUT E., BEYLOT A.-L., "Properties of the MAC layer in safety vehicular *ad hoc* networks", *IEEE Communications Magazine*, vol. 50, no. 11, pp. 1–6, 2012.

[THE 10] THE INSTITUTE OF ELECTRICAL AND ELECTRONICS ENGINEERS, IEEE standard for information technology – telecommunications and information exchange between systems – local and metropolitan area networks – specific requirements. Part 11: wireless LAN Medium Access Control (MAC) and Physical Layer (PHY) specifications – amendment 6: wireless access in vehicular environments, July 2010.

[TOR 09] TORRENT-MORENO M., MITTAG J., SANTI P., *et al.*, "Vehicle-to-vehicle communication: fair transmit power control for safety-critical information", *IEEE Transactions on Vehicular Technology*, vol. 58, no. 7, pp. 3684–3703, 2009.

[WAN 04] WANG C., LI B., LI L., "A new collision resolution mechanism to enhance the performance of IEEE 802.11 DCF", *IEEE Transactions on Vehicular Technology*, vol. 53, no. 4, pp. 1235–1246, 2004.

[WAN 08] WANG Y., AHMED A., KRISHNAMACHARI B., *et al.*, "IEEE 802.11p performance evaluation and protocol enhancement", *Proceedings of the 4th IEEE International Conference on Vehicular Electronics and Safety (ICVES 2008)*, Columbus, pp. 317–322, September 2008.

[WON 06] WONG S., YANG H., LU S., *et al.*, "Robust rate adaptation for 802.11 wireless networks", *Proceedings of the 12th Annual International Conference on Mobile Computing and Networking (MOBICOM 2006)*, Los Angeles, pp. 146–157, September 2006.

[YAN 05] YANG X., VAIDYA N., "On physical carrier sensing in wireless *ad hoc* networks", *Proceedings of the 24th IEEE Annual International Conference on Computer Communications (INFOCOM 2005)*, Miami, pp. 2525–2535, March 2005.

[YAN 07] YANG Y., HOU J., KUNG L., "Modeling the effect of transmit power and physical carrier sense in multi-hop wireless networks", *Proceedings of the IEEE 26th Annual International Conference on Computer Communications (INFOCOM 2007)*, Anchorage, pp. 2331–2335, May 2007.

[YOO 11] YOON Y., KIM H., "Resolving distributed power control anomaly in IEEE 802.11p WAVE", *IEICE Transactions on Communications*, vol. E94-B, no. 1, pp. 290–290, 2011.

[ZHU 04] ZHU J., GUO X., YANG L., *et al.*, "Leveraging spatial reuse in 802.11 mesh networks with enhanced physical carrier sensing", *Proceedings of the 39th IEEE International Conference on Communications (ICC 2004)*, Paris, France, pp. 4004–4011, June 2004.

Chapter 2

Inter-Vehicle Communication for the Next Generation of Intelligent Transport Systems: Trends in Geographic *Ad Hoc* Routing Techniques

2.1. Introduction

Intelligent transport systems (ITS) is an information and communication technology that provides solutions mainly on reducing traffic accidents and congestion. It has been under continuous development since the early 1980s.

Despite there being some early ITS solutions that only adopt autonomous on-board equipment (e.g. ranging sensor and machine vision) to gather information from the surrounding environment, most of the ITS solutions use cooperative approaches, where the traffic-related information is communicated between vehicles and/or

---

Chapter written by Xunxing DIAO, Kun-Mean MOU, Jian-Jin LI and Haiying ZHOU.

roadside infrastructure [PIO 08]. Although an autonomous approach has an advantage in that it does not rely on other participants, it has the obvious limitations in terms of vision and detection ranges as shown in Figure 2.1. A cooperative approach based on single or multiple hops, intervehicle communication (IVC), can better overcome the range limitations and provide a more flexible ITS solution.

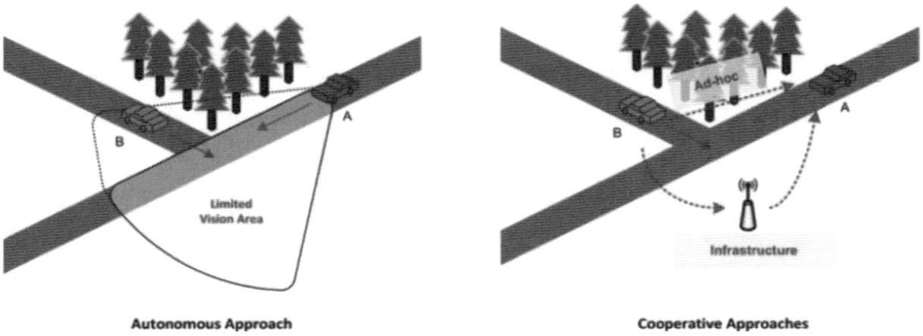

Figure 2.1. *Autonomous approach versus cooperative approaches*

However, the development of an IVC technique is not like developing other civil wireless communication techniques. The IVC has different features as follows:

– An IVC requires more effort to deal with network delay and hard real-time event under a highly dynamic topology. Network delay and system latency cannot be tolerated in some of the ITS applications such as hazard alarming and cooperative driving.

– The size of a vehicular network can be very large in big cities. A traditional client/server system is not likely to be appropriate. The ability to have the distributed configuration is necessary.

– The density of a vehicular network is much more variable. It becomes much lower at night and in bad weather, thus an IVC needs to adapt to the density changes and minimize the administrative overhead.

To develop a real-world IVC technique, these features have to be fully considered. Fortunately, compared with other civil wireless communication techniques, an IVC technique can utilize two extra network supports as follows.

– *Roadside infrastructure support*: there are normally three types of IVC networks: *ad hoc*, infrastructure and hybrid. A vehicular network formed as a pure mobile *ad hoc* network (MANET) is called a vehicular *ad hoc* network (VANET). Since the distribution of a vehicular network is generally along roads, it provides the opportunity for an IVC to access roadside infrastructure and gain extra support on Quality-of-Service (QoS). In a low-density vehicular network, a communication through roadside infrastructures may be the only option.

– *Localization service*: an extra support that an IVC could have is the geographic information (e.g. vehicular position and direction). The information can be provided by the roadside infrastructures and the on-board satellite-based navigation systems. The power supply from on-board battery makes the support practical. The geographic information can enable the geographic Internet protocol (IP) and forwarding strategy that adapt to the frequent changes of network topology with a lower routing overhead and network delay. In fact, to use the geographic information in the development of IVC has become a common assumption.

Due to limited length, this chapter focuses on the geographic routing techniques in a pure VANET, but it does not rule out the possibility of adopting the roadside infrastructure as a localization service. Although this chapter focuses on the routing aspect, it does not leave out the introduction of the techniques in wireless sublayers because they are the foundation of designing any routing technique.

The remainder of the chapter is organized as follows. In section 2.2, the ITS projects related to IVC are summarized.

Then, we introduce the available wireless sublayer techniques that have been developed or experimented by these ITS projects. After that, we present the main sections of the chapter that explains the geographic routing techniques for VANET. In section 2.5, we present the open issues in developing an IVC technique and conclude the chapter.

2.2. IVC-relating ITS projects

The requirements of IVC networks are different in ITS projects. The earlier ITS projects started from enabling the coordinated driving that couples vehicles at close following distances. Normally, the leading vehicle with a human driver works as a base station to perform localized controls. Vehicles use pattern recognition, radar and wireless communication to form a vehicle convoy. Some projects in this period require a localized auto-organized and high-quality real-time IVC system and the typical projects are listed in Table 2.1.

Periods	Projects	Countries	Objectives
1980s	Cooperative driving project by Association of Electronic Technology for Automobile Traffic and Driving (JSK) [TSU 00, SHI 01]	Japan	Group cooperative driving.
1986–2008	California Partners for Advanced Transportation Technology (PATH) Project [CAL 09]	USA	Investigate the control issues of vehicle convoys.
1990s	CHAUFFEUR Project [BEN 99]	EU	Developing new electronic systems for coupling trucks at close following distances.

Table 2.1. *Typical ITS projects in 1980s–1990s*

Most ITS projects in the early 2000s in Table 2.2 tried to provide the IVC-based driver assistance system because it is easier to be implemented and more flexible in complex traffic conditions. The priority of these projects is still safety and traffic efficiency, but the additional goals such as energy efficiency and driving comfort were also considered. Moreover, because the localization technology such as global positioning system/differential global positioning system (GPS/DGPS) has been relatively mature in this stage, the position-based (or called geographic) routing techniques started to be widely used in IVCs.

Periods	Projects	Countries	Objectives
2000–2003	FleetNet Project [FÜß 05]	Germany	Develop a position-based wireless multi-hop *ad hoc* communication platform to improve the driver's and passenger's safety and comfort.
2001–2003	CarTALK 2000 Project [CAR, MOR 03]	EU	– Cooperative driver assistance systems; – Self-organizing *ad hoc* radio network as a communication basis; – Preparing and establishing a future standard.
2001–present	eSafety Working Group [ESA]	EU	Work for a quicker development and increased use of smart road safety and eco-driving technologies (cut road traffic's energy consumption and CO_2 exhaust).

Table 2.2. *Typical ITS projects in the early 2000s*

Following the successes in the previous ITS projects, large ITS projects were starting to be launched in the mid-2000s as shown in Table 2.3. Governments, international commissions and consortiums considerably increases the investment in ITS projects, especially those relating to roadside infrastructure deployment, which is the significant difference compared with the previous period. The existing cellular towers and wireless access points were also considered being utilized as the ITS infrastructure in these projects.

Periods	Projects	Countries	Objectives
2001–present	Projects launched by/related to Car2Car Communication Consortium (C2C-CC) [CARd]	EU	Increasing road traffic safety and efficiency by means of cooperative intelligent transport systems with intervehicle communications supported by vehicle-2-roadside communications.
2003–2011	Driving Safety Support System (DSSS) related projects [INT 09]	Japan	Reduce traffic accidents, relays information to vehicles from roadway infrastructure, advising drivers of nearby traffic lights or warning them of the approach of a vehicle or a pedestrian from a side road.
2003–present	US Department of Transportation (DOT), Vehicle Infrastructure Integration (VII) (The project was renamed to IntelliDrive^{SM} in 2005) [FAR 05, MA 09, CAL 09]	USA	Develop a networked environment supporting very high speed transactions among vehicles (V2V), and between vehicles and infrastructure components (V2I) or hand held devices (V2D) to enable numerous safety and mobility applications.
2004–2010	Smartway Project	Japan	Develop and deploy practical advanced cruise-assist highway system (AHS) infrastructure.
2006–2010	Projects launched by cooperative vehicle-infrastructure system (CVIS) projects launched by European Commission [CVI]	EU	CVISs that create additional effective road network capacity and a more efficient utilization by vehicles.

Table 2.3. *Typical ITS projects after mid-2000s*

Although the ITS is a very active research area, most of the existing ITS projects are still in the planning and design stage. A few experiment platforms have been constructed but there will not be a large implementation until the tested systems are reliable.

The major goals of the current IVC-related ITS projects are to improve safety, traffic management, energy efficiency and driving comfort. A large part of existing IVC techniques are designed based on the assumption that a vehicle can obtain geographic information. The importance of introducing roadside infrastructure in an IVC system has been recognized by recent projects.

2.3. Wireless sublayer techniques

An IVC network connection needs to be made and authenticated very quickly in ITS applications, especially for hazard alarming and cooperative driving. Many of the current wireless techniques at the physical (PHY) layer and the media access control (MAC) layer do not fulfill the high-speed requirement. Therefore, the large ITS projects, particularly those supported by governments and international commissions, have been investigating the related technical issues. Potential solutions, especially for ITS domains, have been proposed and some of them have become standards or obtained patents. The typical PHY and MAC layers techniques that have been tested in real-world experiments are presented in this section.

2.3.1. *WLAN and WPAN (up to 300 m)*

The wireless local area network (WLAN) here only refers to the general Wi-Fi techniques, mainly for personal usage, because technically the dedicated short-range communication (DSRC) discussed in section 2.3.2 is also a type of WLAN technique. On the other hand, the wireless personal area network (WPAN) refers to a more specific personal network for interconnecting devices (e.g. laptops, telephones and Personal Digital Assistants (PDA)) centered on an individual person's workspace. Although many researchers have used the WLAN and WPAN communication models in the IVC protocol designs and simulations, the major techniques that have been tested in real-world environments are only Wi-Fi (IEEE 802.11b [IEE 99]) and ZigBee (IEEE 802.15.4 [SOC 03, SOC 06]):

– Currently, Wi-Fi is the most popular wireless technique. Public Wi-Fi hotspots exist throughout many countries. IEEE 802.11b Wi-Fi has a radio range of up to 100 m and data rate of up to 11 Mbps (For IEEE 802.11g, it is 54 Mbps). The Wi-Fi MAC layer adopts carrier sense multiple access with collision avoidance (CSMA/CA) with an optional request to send/clear to send (RTS/CTS) mechanism to deal with the hidden node problem. The early research at Virginia Tech Transportation Institute in the United States tried to provide a communication solution in high mobility scenarios by using the low-cost WLAN technologies including IEEE 802.11b [AZI 03]. The IntelliDrive™ project used Wi-Fi to assist with DSRC in shorter range communications [CAL 09].

– The IEEE 802.15.4 is a typical low-rate WPAN (LR-WPAN) standard. Compared with Wi-Fi, IEEE 802.15.4 ZigBee has a longer outdoor radio range (up to 300 m) and requires lower power consumption. Moreover, ZigBee devices normally have a lower retail price, which is an important advantage because ITS devices should have a lower price to enable a broader adaptability. The collision avoidance mechanism in 802.15.4 standard is also CSMA/CA. The CIVIC project [DIA 09a, DIA 10] is based on IEEE 802.15.4 and the results from real-world experiments have been published in [DIA 09b].

Both Wi-Fi and ZigBee standards work at unlicensed 2.4 GHz band, thus they eaisly get network interference or management overhead. In addition, their radio ranges are relatively shorter than DSRC discussed in section 2.3.2. However, because these techniques have been widely used in the real world and proved to be generally reliable, the ITS applications built upon them can have a lower implementation cost with a relatively high performance.

Other WPAN and WLAN techniques have been considered for adoption into ITS applications but they all have obvious drawbacks. For example, Bluetooth (IEEE 802.15.1) is also a low-cost standard in 2.4 GHz. It has a 723 Kbps data rate and up to 10 m radio range, but it only supports eight active devices and its network join time is too long (3 s). Similarly, radio-frequency identification (RFID) techniques have problems in terms of radio range and low data rate. They can be used in vehicle identification and electronic toll collection (ETC), but not for general IVC usages.

2.3.2. *Dedicated short-range communication (up to 1 km)*

DSRC is a set of PHY and MAC techniques specifically designed for low-latency ITS solutions. The DSRC-related IEEE standard enables one-way or two-way wireless communication channels in a range up to 1,000 m with a raw data rate of up to 27 Mbps. It works on the licensed frequency band, but the frequency bands allocated for DSRC are not compatible in different countries.

– Early studies on DSRC started in 1993 as the project of a group cooperative driven by the JSK [TSU 00, SHI 01] in Japan. It utilized 5.8 GHz band for transmitting data and used DGPS for measuring vehicle location [TSU 05]. The Japanese national standard of 5.8 GHz DSRC band was completed by the Association of Radio Industries and Businesses (ARIB) in 1995 [OYA 00].

– The Federal Communications Commission (FCC) in the United States established licensing and servic... rules for DSRC in a 75 MHz of bandwidth in 5.9 GHz band... 1999. The leading ITS project IntelliDrive^{SM} [FAR 05, ...] adopts DSRC to support all types of ITS applicatio... 2003, FCC adopted the ASTM E2213-02 [FCC 02] stand... for telecommunications and information exchange betwee... roadside and vehicles.

– In Europe, the European Telecommunications Standards Institute (ETSI) allocated a 30 MHz bandwidth in 5.9 GHz band for ITS in 2008. The main DSRC applications in Europe are in ETC, but currently these applications are still not totally compatible. C2C-CC [CARa] is the major project in Europe to develop the DSRC-related techniques.

Based on the works of FCC ASTM E2213-02 standard, the Institute of Electrical and Electronics Engineers (IEEE) started to create working groups in 2004 to develop the standards to cover the PHY and MAC layers for ITS. The standards and the related techniques are still in the developing stage. The newest results are the IEEE 802.11p [IEE 10] specifications and IEEE 1609 family of standards [WG], which together are called wireless access in vehicular environments (WAVE).

The IEEE 802.11p standard at the MAC layer is an approved amendment to the IEEE 802.11a standard but with lower overhead operations. The MAC layer, frequency band and modulation of these two are very similar. For example, the MAC layer of WAVE also adopts CSMA/CA with an RTS/CTS mechanism. The IEEE 1609 standard family is built on top of IEEE 802.11p to add definitions and more details regarding the architecture, communications model, management structure, security mechanisms and physical access that enable low-latency mobile *ad hoc* and infrastructure vehicular networks.

Name	Max range	Max data rate	Licensed band	Mobility	Infrastructure necessity	Main applications
Bluetooth	10 m	723 Kbps	No	Very low	No	– ETC – Intravehicle usages
RFID	10 m	500 Kbps	No	Low	No	– ETC – Vehicle identification
WLAN: Wi-Fi	100 m (outdoor)	11 or 54 Mbps	No	Middle	No	Low-requirement, low-cost and general-purpose usages
WPAN: ZigBee	300 m (outdoor)	250 Kbps	No	Middle	No	
DSRC	1 km	27 Mbps	Yes	Very high	Customizable	– Safety – Traffic and energy efficiency – Driving comfort
3G: UTRA-TDD	1 km	– 144 Kbps (vehicle) – 384 Kbps (outdoor) – 2 Mbps (indoor).	No	High	Yes	– Traffic and energy efficiency – Driving comfort
4G: WiMAX	50 km	– 40 Mbps (static) – 14 Mbps (mobile)	Customizable	High	Yes	
4G: LTE	100 km	100 Mbps (static and mobile)	Customizable	Very high	Yes	

Table 2.4. *Comparison of considerable wireless sublayer techniques*

2.4. Geographic routing techniques for VANET

This section presents the geographic routing techniques dedicated to the VANET in IVC/ITS applications. First, we discuss the features of VANET by comparing it with the regular MANETs in section 2.4.1. Then, we address one of the main issues in geographic routing, the localization service, in section 2.4.2. Many related chapters just assume this service is already there, but that is not practical in real-world IVC/ITS applications. After that, we present three groups of major geographic routing techniques in VANET including the classical unicast greedy routing in section 2.4.3, the geocast (multicast) routing in section 2.4.4, the delay tolerant network (DTN) based routing in section 2.4.5 and the map-based routing in section 2.4.6.

2.4.1. *Features of VANET*

The VANET is a specific MANET. Compared with the general-purpose MANET, they have three major differences as follows:

– The moving nodes in a VANET have much higher speed. The "mobile" of MANET sometimes just means a random static access in different positions but not really a continuous movement.

– The movement pattern of a VANET is generally limited along the road with fixed directions, but the movement pattern in a MANET is more of a random movement.

– The scale of a VANET is much larger than the scale of a MANET. The short-range sublayer techniques such as Bluetooth, Wi-Fi and ZigBee are all typical MANET techniques.

Because of the low-mobility random pattern and small-scale features, a MANET routing would more often identify nodes by traditional IP methods, and require an end-to-end route to be found before a data delivery. The route discovery normally uses link status as a metric to obtain the best route in a weighted graph. Then, it maintains the weighted graph to adapt to topology changes. But as for a VANET routing, it is a different case. An end-to-end route could often be unavailable in a VANET. The maintenance of traditional IP and route could be very difficult. Simulations [WAN 07, MUS 10] and real-world experiments [WAN 05] also prove the incompatibility of using MANET techniques in a VANET. Therefore, a specialized routing technique needs to be developed.

By considering the VANET and IVC features, and the extra supports from localization service and roadside infrastructure, a geographic routing can be well adapted to ITS applications. The geographic routing techniques consider the physical position of nodes (or regions) as a principle routing parameter. But just to be clear, the geographic routing is not a completely new species in routing techniques. The major difference is just that it utilizes the geographic parameter in the IP and routing processes. Some principles and strategies originally proposed by MANET routing techniques can be transformed into a geographic routing, such as the loop-free mechanism in destination-sequenced distance-vector (DSDV) Routing [PER 94], the reactive and redundancy-free mechanism in *ad hoc* on-demand distance vector (AODV) Routing [PER 97, PER 03], the source routing in dynamic source routing (DSR) [JOH 01], the link-reversal algorithm in temporally-ordered routing algorithm (TORA) [PAR 97] and the zone-based (or cluster-based) mechanism in zone routing protocol (ZRP) [HAA 01]. But in this chapter, these non-geographic routing techniques will not be specifically discussed.

Although the geographic routing techniques have their advantages, they have not yet been become broadly practical in the IVC/ITS projects. Some problems have to be overcome first and these problems are discussed in the following sections.

2.4.2. *Localization*

The first step of a geographic routing is localization, which is one of the main challenges in using geographic routing. Although the satellite-based navigation systems such as GPS (United States), Galileo (European Union) and Beidou (China) are becoming widely available, not all vehicles can have one. Besides, because of the inadequate satellite number (e.g. blocked by buildings), the position may not be always accurate. The localization problems may be improved if a node can gain the supports from the roadside localization service, but some of them could be too expensive to be implemented. Besides, a practical localization service could meet a "the chicken or the egg" dilemma: to improve quality, a practical localization service may require a geographic routing technique to transmit some reference data in the first place. But to transmit the data, a geographic routing may need to acquire the correct geographic information from a localization service in the first place. However, many civil localization services and related techniques have been developed to solve the problems. While some of the localization techniques are relatively mature, it is reasonable to develop the VANET routing techniques based on geographic information.

Geographic routing techniques normally require a networking node to have three types of positions: *its own position*, *the neighbor position* and the *destination position*. Each of them has the related techniques to obtain.

Boukerche *et al.* [BOU 08] summarize the civil localization services that can help a node to obtain its own position.

– *DGPS*: correcting the positions from GPS based on the difference from the positions of reference stations.

– *Map matching*: using the map knowledge to improve positions.

– *Cellular localization*: correcting positions by the mobile cellular infrastructure.

– *Image/video processing*: providing positions through roadside security systems.

– *Indoor infrastructure assist*: using the signal propagation characteristics for indoor environments.

– *Dead reckoning*: calculating the current position based on the last known position.

– *Relative distributed ad hoc localization*: estimating the distance by the known positions of other nodes.

A bigger portion of them relies on the infrastructure supports. The DGPS, video/cam localization and indoor infrastructure assist can provide a more accurate position than others, but they rely on centralized approaches to be realized. The dead reckoning can be independently completed by a node, but it is not accurate for a longer distance.

The positions of neighbor nodes are normally learned through the periodical one-hop broadcast or the reactive neighbor knowledge querying, thus this step is relatively simple. Moreover, if using a contention-based forwarding strategy, there is no need to obtain the neighbor positions in advance.

The main issue here is how to discover the destination position. Normally, the position of the destination node (or region) is specified in the forwarding packets from a source node (original sender). In the best case (e.g. the destination position is a fixed roadside infrastructure), the source node can just obtain the position directly from the roadside infrastructure. In the worst cases, the source node uses the reactive *simple flooding* to query the destination position from all networking nodes that it can reach. Between these two cases, two localization techniques can be adopted, namely *flooding-based localization* and *update and query localization*. Sections 2.4.2.1 and 2.4.2.2 give more details about the two localization techniques, and we summarize them in section 2.4.2.3.

2.4.2.1. *Flooding-based localization*

Distance routing effect algorithm for mobility (DREAM) [BAS 98] represents a typical example of using the proactive flooding-based techniques: a node maintains a position table for the nodes that it can hear, and it tries to send its position-related information to the nodes that it can reach.

To control the localization overhead in flooding, the DREAM protocol considers two effects between nodes: *mobility* and *distance*. The *mobility* effect is implemented as the flooding frequency. The node with a higher speed floods more frequently. The *distance* effect refers to the phenomena that if the distance between two nodes is greater, the relative movement to each other appears to be slower (e.g. for the node A, in effect, the node B seems moving slower than node C in the southerly direction). The packet to deliver the position-related information contains node ID, position, direction and *age* (i.e. hop number). The *age* represents the result of the distance effect. The receivers of such packet can then calculate their distance effect, and decide whether to discard the packet based on the *age* in the packet.

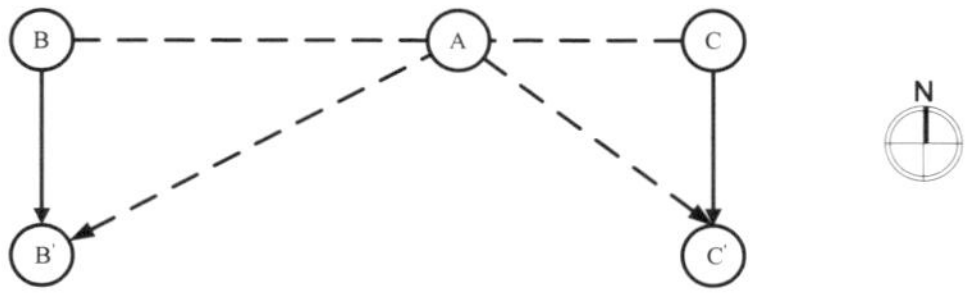

Figure 2.2. *An example to show the distance effect in DREAM*

Another variation of flooding-based localization technique is used in *location-aided routing* (LAR) [KO 00a]. When nodes do not have any knowledge about the network, LAR works similarly to DSR and AODV: reactive request process, avoiding redundant requests in a flooding and the information about route and location is contained in the packets.

Note that both DREAM and LAR only use flooding in the destination discovery, not for the data delivery. More details about their data delivery techniques are given in section 2.4.4.1.

2.4.2.2. *Update and query localization*

There are two major processes in *update and query localization* techniques: *location update* and *destination query*. The former sends out the position-related information to a subset of nodes called the location server; the latter searches the location servers to get a destination location.

The *update and query localization* techniques can be divided into three groups based on the differences in their localization strategies, namely hierarchical [LI 00], quorum-based [LIU 07] and home region [STO 99, RAT 02] localizations. Here, we only introduce the first two techniques, because normally the home region localization is only used in regular MANETs and wireless sensor networks (WSNs).

2.4.2.2.1. Hierarchical localization

The hierarchical localization (or hierarchical hashing-based quorum-based localization) normally explicitly divides nodes into a hierarchical layer structure based on the node positions, and at least a node in each layer acts as a location server that responds to updates and queries for the nodes. The hierarchical localization services can help to reduce the localization overhead and achieve the network scalability, but whether it is robust enough for nodes' mobility like VANETs will need more evaluations to be able to prove. Here, we only introduce a typical protocol called the *grid localization service* (GLS) [LI 00], which has the characters that are suitable for VANETs.

The GLS protocol provides a decentralized hierarchical algorithm, which can handle low-mobility nodes with a lower localization overhead. If all nodes know their GPS positions and they agree on a global origin of the hierarchy as shown in Figure 2.3, the algorithm of GLS can be done by the nodes themselves. Besides, it is possible to further introduce the fuzzy localization into the hashing function, thus not all nodes need to know their accurate GPS positions.

The layer in GLS is referred to as an *order-n* square. A number of *order-n* squares make up an *order-n+1* square as the next layer, and so on. The nodes in the same square must be in each other's one-hop communication distance, and the maximum communication distance is assumed to be two hops. Note that the location update and destination query service does not completely rely on the rules for geographic division.

For the location update (e.g. the node 8), each node periodically delivers its ID to all one-hop neighbors in its first-order square (e.g. to node 20). Then, the location is delivered to the assigned location servers in the next layer (e.g. node 1, 11, 16; maybe delivers from 59 to 16 but it is not important for the algorithm), and the process continues until the ID is delivered to the assigned location servers in all layers (e.g. node 12, 18, 36, then node 9, 10, 53). For each square in the next layers, only one location server will be assigned. The assigned location server is the node with the least ID greater (or greatest ID less) than the ID of the source node; in other words, the node with the closest ID is chosen. For the destination query, it uses a similar process, which tries to find the location server with the closest ID to the destination ID from its layer to the next ones (e.g. node 62 to 12, then 10), and a location server that has stored the ID of the destination will be found eventually.

The GLS protocol optimises the localization overhead by decentralizing the assigning of location servers. Moreover, because the GLS protocol delivers the location update and destination query based on layers, the localization overhead can be greatly reduced and it is predictable: if the height of the hierarchy is $O(log(N))$, effectively the location update and destination query is delivered to $O(log(N))$ location servers, where N is the number of nodes.

2.4.2.2.2. Quorum-based localization

The quorum-based approach means that all nodes in the network agree upon a mapping that maps their unique identifier to one or more quorums. The quorums respond to the specified functions of other nodes.

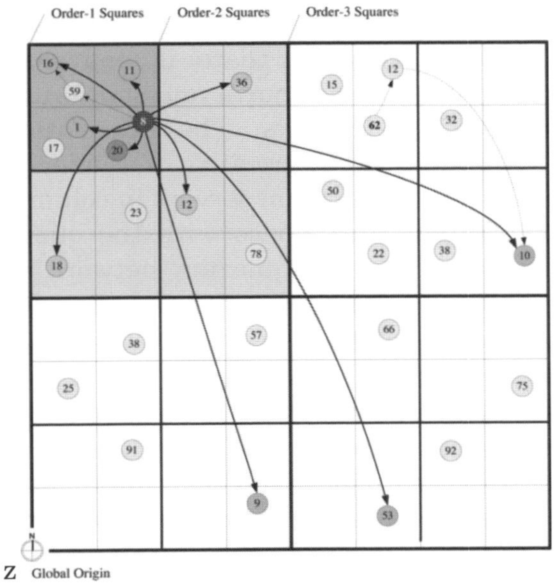

Figure 2.3. *An example of GLS localization*

For quorum-based localization, it normally means that nodes send location updates to a subset of nodes (i.e. location servers), and send destination query to another subset of nodes. These two subsets of nodes must have the intersection nodes to assure a virtual connection backbone. In other cases, if two subsets of nodes are identical, they can also be called *rendezvous based* [FRI 06].

Here, we only introduce the classical quorum-based localization called *column–row* localization such as in dominating set quorum (DS-quorum) [LIU 07] or Column-Row Location Service (XYLS) [DAS 05]. The DS-quorum protocol proposes an algorithm that divides a network into connected dominating sets as shown in Figure 2.4. The dominating set of a graph $G = (V,E)$ is the subset D of V, where the set of vertices in G is either in D or adjacent to a vertex in D. The nodes representing the location servers are arranged in a form of columns and rows, for example the location servers in rows may respond to the location update and the location servers in columns may respond destination query. Then, the location update is delivered from the current location of sender to north and south, until reaching the location servers in rows. The destination query is delivered from the current location of sender to east and west, until crossing the location servers in columns, and then passes to the intersection nodes with the queried location updates. Because the DS-quorum network delivers in column–row form, effectively the location update and destination query are delivered to $O(\sqrt{N})$ location servers.

As for being used in VANETs, there are three advantages of the column–row quorum-based localizations. First, they adapt well to synchronous vehicle movements on roads; second, they can be used to form a network backbone for mixing *ad hoc* and infrastructure communications; third, they are able to better utilize the GPS information about longitudes (columns) and latitudes (rows).

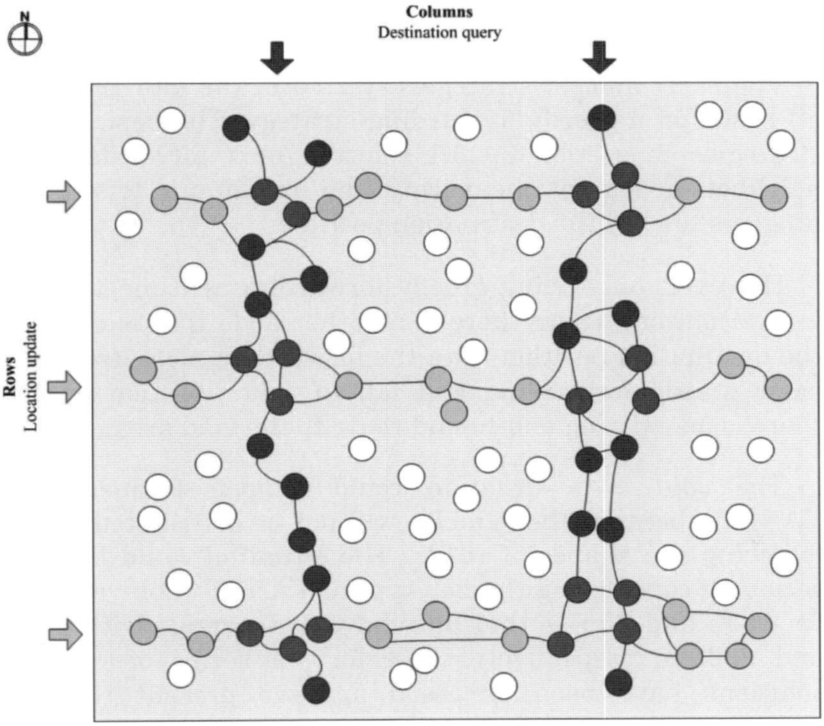

Figure 2.4. *An example of DS-quorum localization*

2.4.2.3. *Summary*

In summary, the flooding-based localization could generate a high localization overhead and does not scale well, but they can have a low implementation complexity, and they are relatively robust in a small network section of a VANET (e.g. the short *ad hoc* sections between cities). On the other hand, the update and query localization can achieve the network scalability which is suitable for a large-scale VANET, but these algorithms themselves may have too much impact on localization overhead, and they are more easily affected by node failures.

2.4.3. *Unicast greedy routing*

The early unicast strategies started in the late 1980s are all based on a greedy forwarding strategy. The basic greedy strategies in section 2.4.3.1 select a next forwarder from neighbor nodes by measuring the maximum forwarding progress toward the destination position.

However, only using greedy forwarding will meet a *void area* situation, where there is no other node that is closer to the destination position than the forwarding node itself. The basic greedy forwarding will fail in this situation even if there is an existing end-to-end route to the void area.

The *void area* situation could happen frequently in VANETs, because the vehicles will not be distributed evenly following the shapes of roads. The situation could be more serious if considering the non-typical VANET sublayers such as Wi-Fi and ZigBee that only have radio range within 100 and 300 m, respectively. Therefore, a series of recovery solutions have been purposed and we present them in section 2.4.3.2.

2.4.3.1. *Basic forwarding strategies*

2.4.3.1.1. Next-hop candidates

Here, we can assume that the positions of a node itself and destination are known from one of the localization services. A geographic greedy routing will then forward a packet to one or more next-hop nodes with the maximum forwarding progress.

A geographic next-hop selection algorithm is normally defined in a *Cartesian* coordinate plane in two dimensions as in Figure 2.5. The network model is assumed to be the unit disk graph where nodes can communicate within radio range R. The node at s is the last sender and the node at d is the destination. From point s to d, it is called *progress direction*.

The area within the radio range and from the y-axis toward the progress direction is called *progress area*. An algorithm can also select the next hop in a smaller progress area, that is the *maximum forwarding area* with a margin in the form of an arc having the center at d. The algorithms are given as follows:

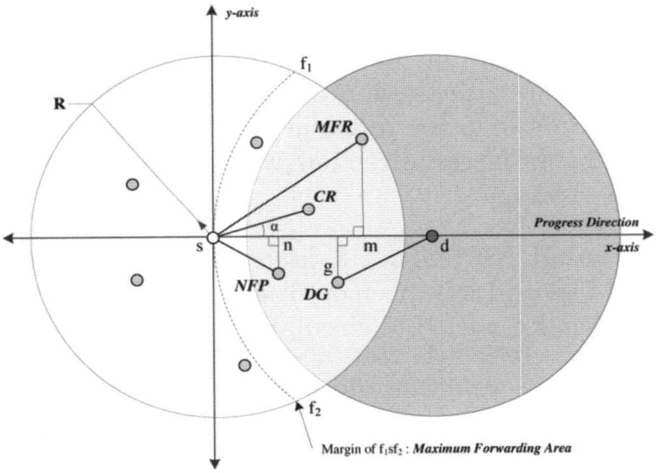

Figure 2.5. *Next-hop candidates in unicast greedy forwarding*

– *Most forward progress within radius (MFR)* [SUI 84]: this strategy selects the node with the longest projection distance in progress direction (e.g. the distance of sm). The longer range transmission is favorable because it may take a packet to more hops ahead. If there is no other node as the next hop to forward a packet, MFR sends the packet back to the previous node.

– *Nearest with forward progress (NFP)* [HOU 86]: the node with the shortest projection distance in the progress area is selected (e.g. the distance of sn). The strategy favors shorter range transmission because it may minimize transmission energy consumption (but it depends on the wireless sublayer functions). Besides, it can have a lower probability of packet collisions if using the contention-based forwarding in section 2.4.3.1.2.

– *Distance-based greedy (DG)* [FIN 87]: the strategy is originally proposed for wired networks. It selects a node that minimizes the distance to the destination (e.g. the distance of *gd*). Its advantage is similar to MFR.

– *Compass routing (CR)* [KRA 99]: it is the first proposal to using the minimum angle in the next-hop selection. It selects the node with the minimum angle between the node and destination (e.g. the angle of α). The nodes closer to the *y*-axis in the progress direction will consume more energy under this strategy.

The original NFP and CR have the problem of routing loop, but MFR and DG are loop free [STO 01]. A routing-loop problem causes a packet to circulate among certain nodes.

2.4.3.1.2. Beacon- and contention-based forwarding

A *beacon-based* forwarding requires knowing the positions of one-hop neighbor nodes, which can be achieved by neighbor knowledge exchanges (i.e. beacon exchanges). When the neighbor positions have been achieved, the selection process for the next-hop node is done by the sender itself. The beacon-based forwarding has less implementation complexity, but it relies on a wireless sublayer to provide a one-hop unicast mechanism, which is available for Wi-Fi, ZigBee and DSRC in section 2.3. The neighbor knowledge exchange could cause additional routing overhead, but it can be reduced if the frequency is well controlled.

A *contention-based* forwarding does not rely on neighbor knowledge exchanges. A sender may blindly broadcast a packet, then the nodes that receive the packet self-configure if they can be the next-hop forwarders. To minimize the packet collision, the number of forwarders needs to be limited by three restrictions as follows:

– The first restriction is that only the nodes in a progress area are selected. An implementation for this restriction is relatively simple. Assuming each node knows its own position, a sender can add its position and the destination position in a forwarding packet. The nodes that receive the forwarding packet can have three required positions and then calculate whether they are the required forwarders.

– The second restriction is to limit next-hop candidates in an area that they can hear from each other, so that if a node in the next-hop candidates has forwarded a packet, the other candidates can know and stop the redundant forwarding.

– If only implementing the previous two restrictions, the node closest to the sender will normally receive and forward the packet first. To avoid that, a geographic forwarding needs to increase the time lag by adding a timer delay function based on the algorithms in section 2.4.3.1.1. For example, if the direct distance d is used, the delay t can be calculated as $t = {}^{a}\!/_{d} \times MaxDelay$, where a is the parameter to adjust the advance progress and *MaxDelay* is the maximum delay to keep a packet before dropping it.

If an RTS/CTS mechanism is available such as the Wi-Fi and DSRC based on IEEE 802.11, the second restriction is optional. A greater involved forwarding area in the progress area can exploit more candidate options such as in [ZOR 03, CHE 07].

For an implementation without an RTS/CTS mechanism such as ZigBee, there are three restriction areas, shown in Figure 2.6, proposed in *beacon-less routing* (BLR) [HEI 04]: a circle with the diameter equal to the radio range R, a *Reuleaux* triangle with the maximum apex angle of 60° or a sector with the same angle. Comparing their proportion with the area of radio range circle one, they can limit the forwarding area to the ratio of about 0.25, 0.22 and 0.17, respectively. A lesser involved forwarding area can reduce the possibility of packet collision. For example, the Implicit Geographic Forwarding (IGF) [BLU 03] based on BLR implements the sector area as the additional RTS/CTS mechanism in IEEE 802.11.

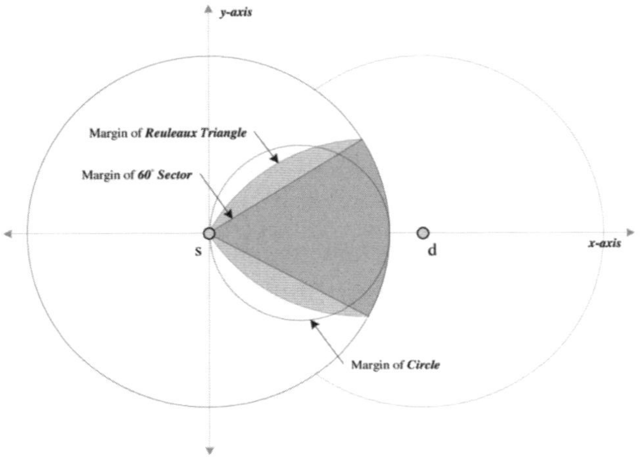

Figure 2.6. *Optional areas in contention-based forwarding*

2.4.3.2. *Void area recovery*

If the progress area of a sender is a void, the forwarding packet will be blocked. The recovery solutions in this case work with the greedy forwarding to deliver the packet. We give more details on the major one, *perimeter routing*, in section 2.4.3.2.1, and then brief the others in section 2.4.3.3.

2.4.3.2.1. Perimeter routing

The *perimeter routing* can provide the best recovery solution. Although its performance relies on an ideal network condition, it can guarantee the packet delivery by only requiring the one-hop neighbor information (if an end-to-end route does exist). Besides, it can work on both beacon-based and contention-based networks.

Planarization

Perimeter routing is a recovery solution based on a planar graph, which is a type of graph with edges that intersect only at their end points. A graph representing a wireless network does not naturally form as a planar graph, thus the graph needs to be simplified by a planarization process. A non-planar graph reduces the performance of a perimeter routing, and it may cause the routing-loop problem [KIM 05, LEE 10a]. The challenge for the planarization in real-world wireless networks is that the nodes can only know the one-hop neighbor information, thus a full planarization for the whole graph is not practical.

Two notable planarization algorithms that require only the one-hop neighbor information are the *Gabriel graph* (GG) [GAB 69] and *relative neighborhood graph* (RNG) [TOU 80]. For both algorithms, if any node x exists within the neighborhood ranges of both A and B (the areas with gray color as shown in Figure 2.7), the edge of (A, B) is removed to avoid the possible crossing edge, and so the remaining edges are (A, x) and (x, B).

Figure 2.7. *Planarization areas of GG and RNG (in gray color)*

GG defines the neighborhood range as a circle with a diameter as the line segment (A, B). RNG defines the neighborhood range as the intersection of two circles with radius R and the circles are centered at A and B. GG and RNG offer different densities of remaining edges (wireless links). RNG produces the planar subgraph with fewer edges, thus it reduces the routing overhead; on the other hand, GG produces the planar subgraph with a better connectivity, thus it may reduce the hop number to a destination.

Face traversal

After the localized planarization process, the nodes obtain a local view of a planar subgraph without edges crossing each other. The next strategy of perimeter routing is to adopt the *right-hand rule* on traversing on the borders of the faces in the planar subgraph. The packets are forwarded face by face and progressively get closer to the destination position.

The first version of the recovery solution using perimeter routing is proposed in [BOS 99], which includes two routing algorithms called *FACE-1* and *FACE-2*. Both FACE-1 and FACE-2 algorithms are not very efficient on their own, but they can guarantee the packet delivery. Thus, they work as the recovery solutions to incorporate with the basic greedy forwarding. Figures 2.8 and 2.9 demonstrate them as the stand-alone routing process without returning to greedy forwarding. The packet in both figures is assumed to be sent from the source node S to the destination node D by a sequence of faces (e.g. from *F1* to *F3*).

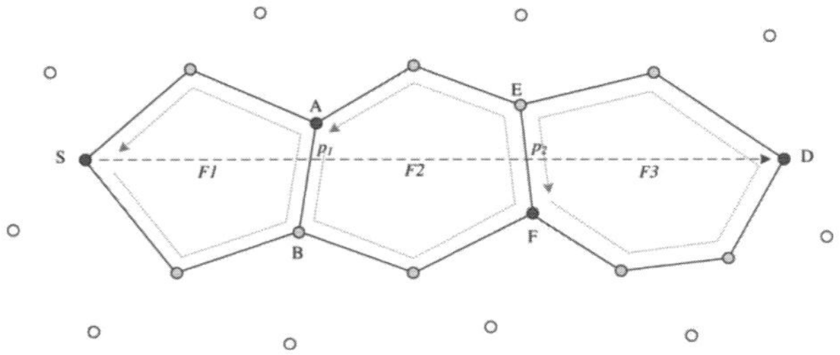

Figure 2.8. *An example of the routing path by FACE 1*

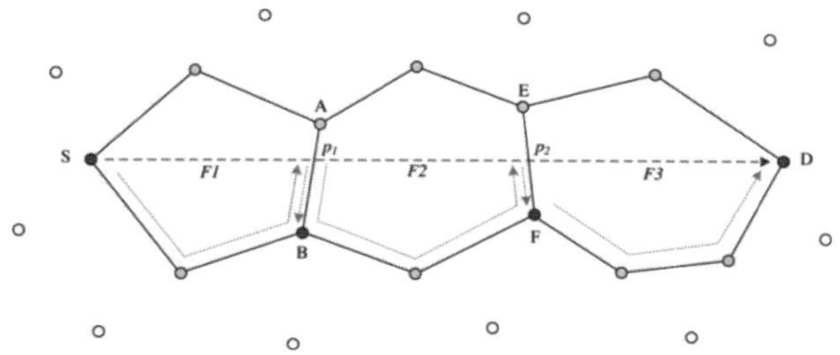

Figure 2.9. *An example of the routing path by FACE-2*

The key rule for *FACE-1* is to find the edges that intersect with the line segment from the source to the destination (e.g. $\overline{SD}$), and the edges found (e.g. (*A, B*) and (*E, F*)) should be closer to the destination gradually (e.g. from *F1* to *F3*, the distances dist (S, D) > dist (p_1, D) > dist (p_2, D)). Before a packet is passed to the next face, the packet must do a complete traversal through the border of a face and then return to the initial point (e.g. *S*, *A* or *F*).

Based on *FACE-1*, *adaptive face routing* (AFR) [KUH 02] purposed a variant algorithm. The source node in AFR initially estimates a boundary of *FACE-1* as an ellipse with foci on source and destination. When a packet reaches the border of the ellipse, the packet is delivered back to the last initial point. The packet is then sent to the initial point of next face. If the routing path is blocked because the ellipse is too small, the packet is sent back to the source node, and the size of the ellipse is increased. If *c* is the cost of the best path in *FACE-1*, AFR can achieve a worst-case cost of $O(c^2)$. Besides, GOAFR+ [KUH 03] purposed an integration of the greedy forwarding and AFR.

FACE-2 is a modified version of *FACE-1*. When a packet is passed to the node with an edge intersecting with the line segment, $\overline{SD}$, the packet is delivered directly to the adjacent face instead of returning to the initial point (e.g. from *B* to *F*, instead of back to *S*). *Greedy-Forward-Greedy* (GFG) is a geographic routing algorithm proposed in [BOS 99], which adopts GG for planarization.

Another well-known beacon-based geographic routing protocol, called *greedy perimeter stateless routing* (GPSR) [KAR 00], implements a recovery solution similar to FACE-2. GPSR proposes the protocol-level details for face routing and an alternative planarization algorithm RNG. When switching faces by GPSR, the packet is always delivered through the first edge of the next face by adopting the right-hand rule. Then, the next edge is searched for in a counterclockwise direction from the last edge. The first edge must be recorded in the transmitting packet until it reaches the next face in order to avoid the routing-loop problem.

Greedy perimeter coordinator routing (GPCR) [LOC 05] is an improved version of GPSR. It utilizes the roads and streets as a communication backbone because they naturally form a planar graph. The greedy and perimeter routing in *GPCR* is only performed when a packet reaches the junctions. Other than that, the packet is forwarded along the road until it reaches the next junction. Therefore, *GPCR* is more efficient than the GPSR in an urban area.

The open issue of the recovery solutions is that they rely too much on an ideal wireless network condition, more precisely, the radio range of these solutions is assumed to be uniform as *R* in a unit disk graph. However, the realistic radio range is often irregular because of the differences in wireless medium densities, link errors and inaccurate positions. Some solutions were proposed for the non-ideal network conditions, for example Cross Link Detection Protocol (CLDP) [KIM 05] uses an additional proactive message for planarization and Greedy Distributed Spanning Tree Routing (GDSTR) [LEO 06] uses the traversal of a *hull spanning tree* (an alternative technique of planarization). However, while the former increases the routing overhead significantly, the latter loses the localizable advantage in geographic routing.

2.4.3.3. *Other than recovery*

The techniques in this section can be used if a void area recovery is unavailable. There are three groups of solutions including dropping a blocked packet, sending it back or exploiting more hops in advance.

Dropping the blocked packets can be an option only if (1) the nodes are generally moving and a resend mechanism is available or (2) a multipath routing is already used so the packet is supposed to reach a destination on the other path. SPEED [HE 03] is a beacon-based solution, which considers dropping the blacked packet to reduce the traffic congestion. Each node in SPEED records the average delays to destinations in its neighboring table. When meeting a void area, the delay is marked as ∞. The neighbors then receive the notice for the void area by the so-called *backpressure beacon*.

Another suggestion is to send a blocked packet back to the last forwarder. The failed routing path will be marked, thus the new greedy forwarding can look for another path and avoid a routing loop. If the mobility of nodes is considered, any node in a similar position to that of the last forwarder can be used as a backtracking node. Furthermore, GDSTR [LEO 06] maintains a spanning tree where each node has an associated *convex hull* that contains within it the locations of all its descendant nodes in the tree. When a void area is found, the block packets are routed upwards in the tree until a node whose convex hull contains the destination is found.

In a beacon-based forwarding, a solution is to exploit more hops neighbor information in advance. The result in [STO 01] shows that if two-hop geographic information such as GEDIR, DIR and MFR is available for each node, the void area problem can be reduced. The trade-off for the two-hop geographic information is an additional routing overhead.

2.4.4. *Geocast (multicast) routing*

The geocast forwarding steps are similar to the contention-based forwarding, but the destination in geocast is more often a geographic cluster. If the destination is only a single node, when packets reach the border of the cluster that contains the destination node, the transmitting mode can be switched back to the unicast mode.

Moreover, the geocast forwarding steps can be assisted by two other techniques: *hierarchy* and *flooding*. The hierarchical geocast (e.g. GeoTora [KO 00b] and GeoNode [IMI 99]) forward packets cluster by cluster, thus it can reduce routing overhead and increase network scalability. However, the trade-off of these advantages is an overhead in cluster division. For a small area IVC built on IEEE 802.11-based Wi-Fi or DSRC, the cluster division could be too short-lived to be worth creating. The hierarchical geocast may only be suitable for a large area IVC based on 3G or 4G.

The following sections only describe the non-hierarchy flooding-based geocast techniques for VANETs. In this context, the geocast applications are only for distributing emergency messages, for example to deliver a collision warning to all approaching vehicles and nearby junctions. In the following, we will introduce two typical flooding-based geocast techniques in section 2.4.4.1 and then the related geocast techniques for VANETs in section 2.4.4.2.

2.4.4.1. *Restricted directional flooding*

DREAM [BAS 98] and LAR [KO 00a] are two broadly adopted geocast protocols. They both adopt the restricted directional flooding in their data transmission, but their restricted areas are different.

After the localization steps of DREAM and LAR discussed in section 2.4.2.1, assuming a source node S in DREAM or LAR know that the destination node D is in the position (x_d, y_d) at time t_0, and that the current time is t_1, the node can then restrict the direction and area of the next flooding as shown in Figure 2.10. The key scheme for both protocols is to ensure that a packet is sent to an expected region and that the destination node will be there when the packet reaches the expected region.

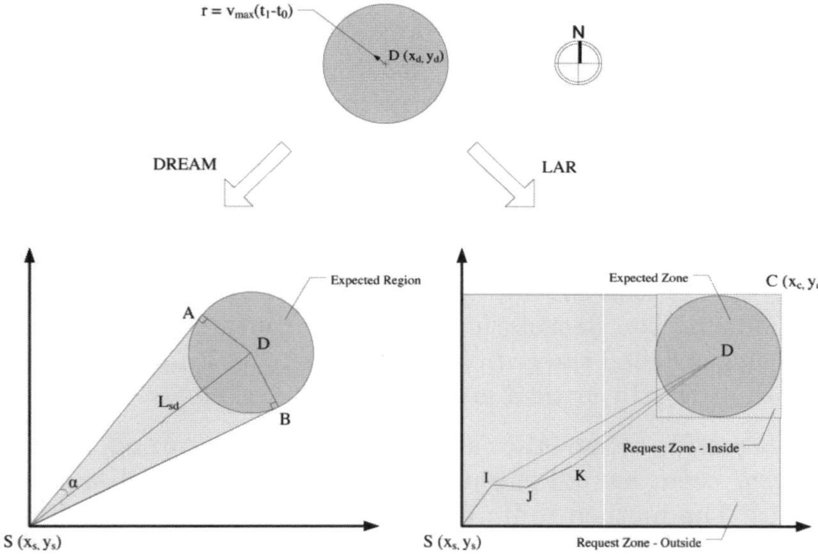

Figure 2.10. *Flooding areas in DREAM and LAR*

Both DREAM and LAR expect the node D to be in the circle area centered at (x_d, y_d) with the radius of $r = v_{max}(t_1 - t_0)$ (e.g. the *expected region* (*zone*) are the same circle areas in the northeast from node S), but the next steps are different.

– For DREAM, the nodes involved in the flooding process are the nodes within the forwarding angle α defined as $\alpha = arcsin\ (r / L_{sd})$, where L_{sd} is the distance between nodes S and D.

– The LAR (scheme I) defines a *request zone* as a rectangle, where only the nodes in the rectangle floods the data packets. If a nodes S is outside the expected zone, the rectangle is within (x_S, y_S) and (x_C, y_C). If a node S is inside the expected zone, the minimum boundary of the request zone cannot be smaller than the expected zone. The LAR (scheme II) further defines that only the nodes with the shorter distances to the destination node can be involved in the next-hop flooding process (e.g. from the node I, J to E).

2.4.4.2. *Flooding-based geocast for VANETs*

Two earlier examples of flooding-based VANET geocast protocols are the geocast scheme in [BRI 00] and the IVG in [BAC 03]. The basic strategies of them are similar.

– First, when an accident happens, an alarm message will be sent out to all the vehicles that will be affected by the accident. For example, if the accident is on a roundabout, only the vehicles driving toward the roundabout will receive the alarm. The destination area that contains the alarmed vehicles is called the *critical area*. If vehicles know their GPS information, the critical area can be defined easily.

– Second, when the alarm message is spread in the *critical area*, not all the nodes need to be involved as relay nodes. The method to limit the number of relay nodes is the same method as in contention-based forwarding (in section 2.4.3.1.2).

In addition, there are other geocast algorithms and protocols for VANETs similar to these two earlier examples but with unique features.

– *Cached geocast* in [MAI 04] proposes including caching at the routing layer to deal with the situation of high velocities in VANETs. The small cache can help to improve the problems of neighbor selection and void area in the geocast forwarding.

– *Urban multi-hop broadcast* (UMB) [KOR 04] redefines the RTS/CTS mechanism in the IEEE 802.11 standard to address the problems of broadcast storm, hidden node and reliability problems of multi-hop broadcast in urban areas. The UMB divides the road into segments in the direction of dissemination, and only one vehicle in each segment has the duty of forwarding and acknowledging the packets.

– *Abiding geocast* [MAI 05] is a specific geocast technique. The abiding geocast can be used to send messages to a fixed geographical area (e.g. the warning of an icy road in winter). Besides the regular approach such as the periodical delivery, *abiding geocast* provides three more options: a server approach, a node election in the destination region and a neighbor exchange solution.

2.4.5. *Delay tolerant network-based routing*

DTN is an extreme case of MANET. VANET can be treated as a form of DTN. The distinguished feature of DTN is that the end-to-end connectivity between the source and destination in DTN is assumed to be frequently broken due to network partitioning.

The earliest research on DTN routing mostly uses the flooding-based techniques, but a more recent research direction tries to utilize the movement feature of nodes instead of adapting to it. That is why the recent DTN techniques are very suitable for VANETs.

This section provides two interesting DTN-based routing options that utilize the movement feature in VANETs: *last encounter routing* (LER) [GRO 03, GRO 06] and *carry-and-forward routing* [DAV 01].

2.4.5.1. *Last encounter routing*

An example of LER is a routing algorithm called *exponential age search* (EASE) [GRO 03, GRO 06]. The recent application of LER is in the FleetNet project, which tries to build a virtual flea market over VANET. The customers express their demands/offers by smart phones, PDAs and laptops within a VANET.

Grossglauser and Vetterli [GRO 03] first proposed a movement-based localization service, and it shows that it is possible to only use the node mobility to disseminate destination location information without using any flooding-based method. In other words, only "free" information about the local connectivity to neighboring nodes is adopted. Then, a simple routing algorithm called EASE was proposed to evaluate such localization service. The interesting conclusion about EASE is that the collections of last encounter histories at network nodes contain enough information for a geographic routing protocol to route packets.

For the part of localization service, each node in EASE maintains a *last encounter table* (LET), which contains three fields including *node ID*, *location* and *time*. If a node i meets a node j at position P_{ij}, node i records an entry as *node ID* equaling j and *location* equaling P_{ij}. *Time* for the entry is the time elapsed since the encounter at P_{ij}.

As for the routing part, the principal steps are as follows: when a source node tries to send a packet, the source node search its neighbors until finding a neighbor who meets the destination in the latest time based on the information of LET. Then, the packet is routed toward the latest encounter location. The process continues until the packet reaches the destination node. For example, vehicle S tries to send a packet to vehicle A as shown in Figure 2.11. In its current radio range, vehicle B meets vehicle A at the location of B_2. If the location B_2 available on B is newer than any other location information that the vehicle S can obtain, the packet is sent to the location B_2. The EASE made no assumptions about how to route the packet toward a latest encounter location, and any geographic routing protocol can be used here. The disadvantage of EASE is the delivery fails easily in a practical network when the network just starts up, or where there is limited radio range, thus the number of neighbors is too small.

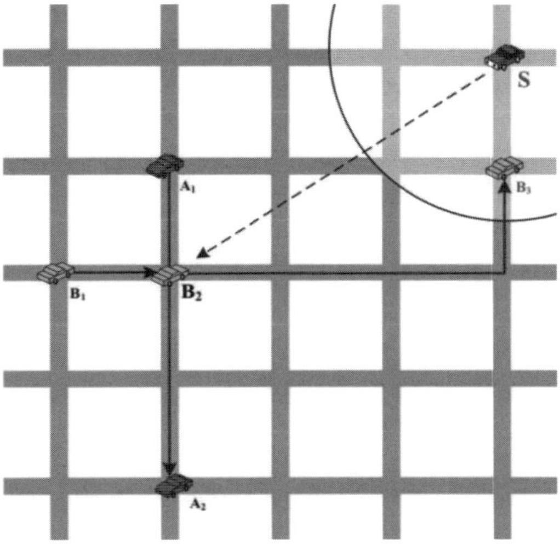

Figure 2.11. *An example of the routing path by EASE*

2.4.5.2. *Carry-and-forward routing*

Carry-and-forward is a new concept proposed in [DAV 01]. The idea is as the name suggests: when a routing path does not exist for a packet, the last receiver can carry the packet and forward the packet to the new receiver until some conditions meet.

An example protocol adopting the carry-and-forward concept is *vehicle-assisted data delivery* (VADD) [ZHA 06]. A moving vehicle in VADD carries a packet and forwards it to the next vehicle in the intersection of roads. In other words, the routing paths in VADD are the exact shape of the roads. Moreover, VADD predicts the mobility of other vehicles, which follows the traffic pattern and road layout. A routing decision is based on the result of such a prediction. The experimented routing decisions are based on location (*L-VADD*), direction (*D-VADD*), multipath direction (*MD-VADD*) and hybrid (*H-VADD*). The *H-VADD* protocol has much better performance and it can avoid the routing-loop problem.

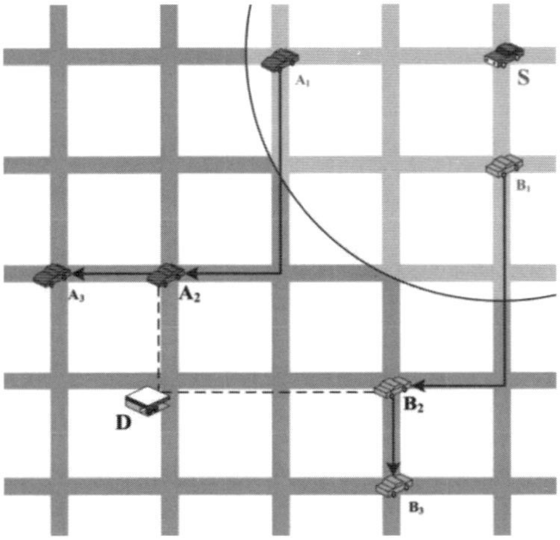

Figure 2.12. *An example of the routing path by GeOpps*

Geographical opportunistic routing (GeOpps) [LEO 07] is another carry-and-forward protocol, which requires navigation information of other vehicles to predict the mobility of other vehicles. By knowing the navigation information, the node in GeOpps knows the paths of other vehicles when it tries to forward a packet, then a decision can be made by comparing the nearest point of these path to the destination point. For example, vehicle S in Figure 2.12 tries to find a routing path to the gas station at D. Two vehicles, A and B, are in the radio range of S, and they will be driven from A_1 to A_3 and from B_1 to B_3, respectively. The nearest point of these two routing path is A_2, thus A becomes the new relay in the routing path. The GeOpps in theory can get a better result than VADD, but the navigation information utilized in GeOpps is mostly private in VANETs.

2.4.6. *Map-based routing*

Maybe we can call the map-based routing a semi-geographic VANET technique. This technique does not directly use the hop-to-hop querying and forwarding as in previous sections; instead, it uses a global road map provided by roadside infrastructure for calculating the shortest path. Because the map-based routing relies heavily on the supports from roadside infrastructure, it is not really a pure VANET routing technique. Here, we provide a brief because this technique can be very practical in a metropolis area. Moreover, the simulation results [LEE 10b, MUS 10] show that the map-based technique can significantly outperform the techniques without using road map.

Geographic source routing (GSR) [LOC 03] is a typical example. It uses a *reactive location service* (RLS), which has some similarity with the localization service in DREAM but in a reactive approach, to obtain the destination position. Then, it calculates the junctions in the road map that will be used in traversal by using the Dijkstra's shortest path algorithm in a weighted graph, where the vertices are junctions and the edges are streets. The forwarding process between junctions is position-based.

Because the GSR only uses a static road map for its calculation, the obtained route may be a road without enough passing vehicles to be forwarding nodes. *Anchor-based street and traffic aware routing* (A-STAR) [LIU 04] improves the GSR by adding a traffic awareness process, which utilizes the city bus paths as an overlay map in order to identify the truck roads with higher connectivity.

2.5. Conclusion and open issues

It is still ongoing work for many research projects to develop a reliable IVC for ITS applications to fulfill the requirements of safety, traffic management, energy efficiency and driving comfort. It is really a difficult task because of three unique features of vehicular network: high mobility, large scale and variable density. Great research efforts have been put in this area by the schools, governments and consortiums. Although there are many potential solutions, the practical solution has not yet been found.

This chapter is mainly about the routing techniques in IVC/ITS applications, and we focus more on introducing the practical VANET routing techniques that can be used currently or in the near future. The wireless sublayer techniques on PHY and MAC are also presented in this chapter, because they are the foundations for building any practical routing technique.

The main body of the chapter gives a comprehensive survey on variant geographic routing techniques, because by considering the development of localization services, the geographic routing is quite clearly the best suitable solution for IVC/ITS applications. The survey presents the main research direction in geographic routing techniques, the unicast greedy routing, with three additional or substitutable techniques including geocast, DTN-based and map-based techniques.

The chapter has introduced the main open issues and new techniques in these geographic routing techniques along with the technique in localization service, but there are still many of them that have not been fully addressed, such as the security problem in IVC/ITS, the conversion between Internet Protocol version 4 (IPv4)/Internet Protocol version 6 (IPv6) and geographic position, the location-aware transport layer techniques and the QoS problems in low-cost sublayer techniques. The research outcomes of these areas will surely improve the reliability and efficiency in IVC/ITS applications.

2.6. Acknowledgments

This work was sponsored by three organizations: the French government through the research program "Investissements d'avenir" at the IMobS3 Laboratory of Excellence (ANR-10-LABX-16-01), the European Union through the program "Regional Competitiveness and Employment" 2007–2013 (ERDF–Auvergne region), and the Auvergne region.

2.7. Bibliography

[AZI 03] AZIZ F.M., Implementation and Analysis of Wireless Local Area Networks for High-Mobility Telematics, Virginia Polytechnic Institute and State University, 2003.

[BAC 03] BACHIR A., BENSLIMANE A., "A multicast protocol in ad hoc networks inter-vehicle geocast", *Proceedings of the 57th IEEE Semiannual Vehicular Technology Conference*, Jeju, South Korea, 2003.

[BAS 98] BASAGNI S., *et al.*, "A distance routing effect algorithm for mobility (DREAM)", *Proceedings of the 4th Annual ACM/IEEE International Conference on Mobile Computing and Networking*, ACM, New York, NY, 1998.

[BEN 99] BENZ T., *et al.,* CHAUFFEUR – TR 1009 – User, Safety and Operational Requirements, Community Research and Development Information Service (CORDIS), 1999.

[BLU 03] BLUM B.M., *et al.*, IGF: A State-Free Robust Communication Protocol for Wireless Sensor Networks, Computer Science Department, University of Virginia, 2003.

[BOS 99] BOSE P., *et al.*, "Routing with guaranteed delivery in ad hoc wireless networks", *Proceedings of the 3rd International Workshop on Discrete Algorithms and Methods for Mobile Computing and Communications*, ACM, New York, NY, 1999.

[BOU 08] BOUKERCHE A., *et al.*, "Vehicular ad hoc networks: a new challenge for localization-based systems", *Computer Communications*, vol. 31, no. 12, pp. 2838–2849, 2008.

[BRI 00] BRIESEMEISTER L., *et al.*, "Disseminating messages among highly mobile hosts based on inter-vehicle communication", *IEEE Intelligent Vehicles Symposium*, Dearborn, Michigan, USA, 2000.

[CAL 09] California PATH Annual Report, 2009. Available at http://www.path.berkeley.edu/.

[CARa] CAR 2 CAR Communication Consortium (Car2Car CC), 2013. Available at http://www.car-to-car.org/.

[CARb] CarTALK2000 Website, 2001. Available at http://www.cartalk2000.net/.

[CHE 07] CHEN D., DENG J., VARSHNEY P.K., "Selection of a forwarding area for contention-based geographic forwarding in wireless multi-hop networks", *IEEE Transactions on Vehicular Technology*, vol. 56, no. 5, pp. 3111–3122, 2007.

[CVI] CVIS Project Website, 2009. Available at http://www.cvisproject.org/.

[DAS 05] DAS S.M., PUCHA H., HU Y.C., "Performance comparison of scalable location services for geographic ad hoc routing", *INFOCOM 2005, 24th Annual Joint Conference of the IEEE Computer and Communications Societies, Proceedings IEEE*, Miami, USA, vol. 2, pp. 1228–1239, 2005.

[DAV 01] DAVIS J.A., FAGG A.H., LEVINE B.N., "Wearable computers as packet transport mechanisms in highly-partitioned ad-hoc networks", *Proceedings of the 5th IEEE International Symposium on Wearable Computers*, IEEE Computer Society, Washington, DC, 2001.

[DIA 09a] DIAO X.X., *et al.*, "Cooperative inter-vehicle communication protocol with low cost differential GPS", *Journal of Networks*, vol. 4, no. 6, pp. 445–457, 2009.

[DIA 09b] DIAO X., *et al.*, "Experiments on PAVIN platform for cooperative inter-vehicle communication protocol (CIVIC)", *IEEE Conference in Africa (AFRICON)*, Kenya, 2009.

[DIA 10] DIAO X.X., *et al.*, "An embedded system dedicated to intervehicle communication applications", *Hindawi Publishing Corporation EURASIP Journal on Embedded Systems*, p. 15, 2010.

[ESA] eSafety Website, 2012. Available at http://ec.europa.eu/information_society/activities/esafety/.

[FAR 05] FARRADYNE P., Vehicle Infrastructure Integration (VII) Architecture and Functional Requirements, ITS Joint Program Office, 2005.

[FCC 02] FCC, ASTM E2213-02 Standard Specification for Telecommunications and Information Exchange Between Roadside and Vehicle Systems — 5 GHz Band Dedicated Short Range Communications (DSRC) Medium Access Control (MAC) and Physical Layer (PHY) Specifications, Federal Communications Commission (FCC), 2002.

[FIN 87] FINN G.G., *Routing and Addressing Problems in Large Metropolitan-Scale Internetworks*, University of Southern California, Information Sciences Institute, 1987.

[FRI 06] FRIEDMAN R., KLIOT G., *Location Services in Wireless Ad Hoc and Hybrid Networks: A Survey*, Department of Computer Science Technion, Israel, 2006.

[FÜß 05] FÜ LER H., *et al.*, "Position-based routing in ad-hoc wireless networks", in FRANZ W., HARTENSTEIN H., MAUVE M. (eds), *Inter-Vehicle-Communications Based on Ad Hoc Networking Principles - The FleetNet Project*, Göttingen University Press, Karlsruhe, Germany. pp. 117–143, 2005.

[GAB 69] GABRIEL R.K., SOKAL R.R., "A new statistical approach to geographic variation analysis", *Systematic Zoology*, vol. 18, no. 3, pp. 259–278, 1969.

[GRO 03] GROSSGLAUSER M., VETTERLI M., "Locating nodes with EASE: last encounter routing in ad hoc networks through mobility diffusion", *INFOCOM 2003, Twenty-Second Annual Joint Conference of the IEEE Computer and Communications, IEEE Societies*, San Francisco vol. 3, pp. 1954–1964, 2003.

[GRO 06] GROSSGLAUSER M., VETTERLI M., "Locating mobile nodes with EASE: learning efficient routes from encounter histories alone", *IEEE/ACM Transactions on Networking*, vol. 14, no. 3, pp. 457–469, 2006.

[HAA 01] HAAS Z.J., PEARLMAN M.R., "The performance of query control schemes for the zone routing protocol", *IEEE/ACM Transactions on Networking*, vol. 9, no. 4, pp. 427–438, 2001.

[HE 03] HE T., *et al.*, "SPEED: a stateless protocol for real-time communication in sensor networks", *Proceedings of the 23rd International Conference on Distributed Computing Systems*, Providence, RI, USA, 2003.

[HEI 04] HEISSENBÜTTEL M., BRAUN T., BERNOULLI T., *et al.*, "BLR: beacon-less routing algorithm for mobile ad hoc networks", *Computer Communications*, vol. 27, no. 11, pp. 1076–1086, 2004.

[HOU 86] HOU T.-C., LI V., "Transmission range control in multihop packet radio networks", *IEEE Transactions on Communications*, vol. 34, no. 1, pp. 38–44, 1986.

[IEE 99] IEEE, *Part 11: Wireless LAN Medium Access Control (MAC) and Physical Layer (PHY) specifications: Higher-Speed Physical Layer Extension in the 2.4 GHz Band*, IEEE Computer Society, 1999 (Retrieved 2007).

[IEE 10] IEEE Standard for information technology – Telecommunications and information exchange between systems – Local and metropolitan area networks – Specific requirements Part 11: Wireless LAN Medium Access Control (MAC) and Physical Layer (PHY) Specifications – Amendment 6: Wireless Access in Vehicular Environments, 2010. Available at http://standards.ieee.org/getieee802/download/802.11p-2010.pdf.

[IMI 99] IMIELIŃSKI T., NAVAS J.C., "GPS-based geographic addressing, routing, and resource discovery", *Communications of the ACM*, vol. 42, pp. 86–92, 1999.

[INT] Introductions of Driving Safety Support System (DSSS) Project, 2009. Available at http://www2.toyota.co.jp/en/news/11/06/ 0629.html; http://global-sei.com/sn/2009/384/3.html.

[JOH 01] JOHNSON D.B., MALTZ D.A., BROCH J., "DSR: the dynamic source routing protocol for multi-hop wireless ad hoc networks", in PERKINS C.E. (ed.), *Ad Hoc Networking,* Chapter 5, Addison-Wesley, pp. 139–172, 2001.

[KAN 09] KANDARPA R., *et al.*, Final Report: Vehicle Infrastructure Integration Proof-of-Concept Technical Description – Infrastructure, Research and Innovative Technology Administration (RITA) U.S. Department of Transportation, 2009.

[KAR 00] KARP B., KUNG H.T., "GPSR: greedy perimeter stateless routing for wireless networks", *Proceedings of the 6th Annual International Conference on Mobile Computing and Networking (MobiCom '00)*, ACM, New York, NY, 2000.

[KIM 05] KIM Y.-J., *et al.*, "Geographic routing made practical", *Proceedings of the 2nd Conference on Symposium on Networked Systems Design & Implementation*, USENIX Association, Berkeley, CA, vol. 2, 2005.

[KO 00a] KO Y.-B., VAIDYA N.H., "Location-aided routing (LAR) in mobile ad hoc networks", *Wireless Networks*, vol. 6, no. 4, pp. 307–321, 2000.

[KO 00b] KO Y.-B., VAIDYA N.H., "GeoTORA: a protocol for geocasting in mobile ad hoc networks", *Proceedings of the 2000 International Conference on Network Protocols*, IEEE Computer Society, Washington, DC, 2000.

[KOR 04] KORKMAZ G., ÖZGÜNER F., EKICI E., *et al.*, "Urban multi-hop broadcast protocol for inter-vehicle communication systems", *Proceedings of the 1st ACM International Workshop on Vehicular Ad Hoc Networks (VANET '04),* ACM, New York, NY, 2004.

[KRA 99] KRANAKIS E., SINGH H., URRUTIA J., "Compass routing on geometric networks", *11th Canadian Conference on Computational Geometry*, Vancouver, Canada 1999.

[KUH 02] KUHN F., WATTENHOFER R., ZOLLINGER A., "Asymptotically optimal geometric mobile ad-hoc routing", Proceedigs of the 6th International Workshop on Discrete Algorithms and Methods for Mobile Computing and Communications, ACM, New York, NY, 2002.

[KUH 03] KUHN F., WATTENHOFER R., ZOLLINGER A., "Worst-case optimal and average-case efficient geometric ad-hoc routing", *Proceedings of the 4th ACM International Symposium on Mobile Ad Hoc Networking & Computing*, ACM, New York, NY, 2003.

[LEE 10a] LEE K.C., CHENG P.-C., GERLA M., "GeoCross: a geographic routing protocol in the presence of loops in urban scenarios", *Ad Hoc Networks*, vol. 8, pp. 474–488, 2010.

[LEE 10b] LEE K.C., LEE U., GERLA M., "Survey of Routing Protocols in Vehicular Ad Hoc Networks", in *Advances in Vehicular Ad-Hoc Networks: Developments and Challenges*, IGI Global, 2010.

[LEO 06] LEONG B., LISKOV B., MORRIS R., "Geographic routing without planarization", *Proceedings of the 3rd Conference on 3rd Symposium on Networked Systems Design & Implementation (NSDI '06)*, USENIX Association, Berkeley, CA, 2006.

[LEO 07] LEONTIADIS I., MASCOLO C., "GeOpps: geographical opportunistic routing for vehicular networks", *World of Wireless, Mobile and Multimedia Networks, 2007 (WoWMoM 2007), IEEE International Symposium on a World of Wireless, Mobile and Multimedia Networks*, Helsinki, Finland, 2007.

[LI 00] LI J., *et al.*, "A scalable location service for geographic ad hoc routing", *Proceedings of the 6th Annual International Conference on Mobile Computing and Networking (MobiCom '00)*, ACM, New York, NY, 2000.

[LIU 04] LIU G., *et al.*, "A routing strategy for metropolis vehicular communications information networking. Networking technologies for broadband and mobile networks", in KAHNG H.-K., GOTO S., (eds), *Networking Technologies for Broadband and Mobile Networks*, Berlin/Heidelberg, Springer, pp. 134–143, 2004.

[LIU 07] LIU D., JIA X., STOJMENOVIC I., "Quorum and connected dominating sets based location service in wireless ad hoc, sensor and actuator networks", *Computer Communications*, vol. 30, pp. 3627–3643, 2007.

[LOC 05] LOCHERT C., *et al.*, "Geographic routing in city scenarios", *SIGMOBILE Mobile Computing and Communications Review*, vol. 9, no. 1, pp. 69–72, 2005.

[LOC 03] LOCHERT C., *et al.*, "A routing strategy for vehicular ad hoc networks in city environments", *CORD Conference Proceedings*, Columbus, OH, USA, pp. 156–161, 2003.

[MA 09] MA Y., *et al.*, "Real-time highway traffic condition assessment framework using vehicle-infrastructure integration (VII) with artificial intelligence (AI)", *Transactions on Intelligent Transportation Systems*, vol. 10, pp. 615–627, 2009.

[MAI 04] MEIHÖFER C., EBERHARDT R., "Geocast in vehicular environments: caching and transmission range control for improved efficiency", *IEEE Intelligent Vehicles Symposium (IV)*, Parma, Italy, 2004.

[MAI 05] MAIHÖ F., CHRISTIAN L. LEINMÜ, TIM, SCHOCH E., "Abiding geocast: time–stable geocast for ad hoc networks", *Proceedings of the 2nd ACM International Workshop on Vehicular Ad Hoc Networks (VANET '05)*, ACM, New York, NY, 2005.

[MOR 03] MORSINK P., *et al.*, "CarTALK2000: development of a co-operative ADAS based on vehicle-to-vehicle communication", *10th World Congress and Exhibition on Intelligent Transport Systems and Services*, Madrid, Spain, 2003.

[MUS 10] MUSTAFA B., RAJA U.W., "Issues of routing in VANET", School of Computing, Blekinge Institute of Technology, 2010.

[OYA 00] OYAMA S., TACHIKAWA K., SATO M. "DSRC standards and etc systems development in Japan", *Proceedings of the 7th World Congress on Intelligent Systems*, Turin, Italy, 2000.

[PAR 97] PERKINS, CHARLES E. and BHAGWAT, PRAVIN, "Highly dynamic Destination-Sequenced Distance–Vector routing (DSDV) for mobile computers", *Proceedings of the Conference on Communications Architectures, Protocols and Applications*, ACM, London, United Kingdom, pp. 234–244, 1994.

[PER 94] PERKINS C.E., BHAGWAT P., "Highly dynamic destination-sequenced distance-vector routing (DSDV) for mobile computers", *Proceedings of the conference on Communications architectures, protocols and applications*, ACM, London, United Kingdom, pp. 234–244, 1994.

[PER 97] PERKINS C.E., Ad Hoc On-Demand Distance Vector Routing Protocol, IETF MANET Working Group, 1997.

[PER 03] PERKINS C., ROYER E., DAS S., Ad hoc On-Demand Distance Vector (AODV) Routing (Retrieved 18 May 2010), The Internet Society, 2003.

[PIO 08] PIAO J., MCDONALD M., "Advanced driver sssistance systems from autonomous to cooperative approach", *Transport Reviews*, vol. 28, no. 5, pp. 659–684, 2008.

[RAT 02] RATNASAMY S., *et al.*, "GHT: a geographic hash table for data-centric storage", *Proceedings of the 1st ACM International Workshop on Wireless Sensor Networks and Applications*, ACM, New York, NY, 2002.

[SHI 01] SHIRAKI Y., *et al.*, "Development of an inter-vehicle communications system", *Special Edition on ITS*, vol. 68, pp. 11–13, 2001.

[SMA] Smartway WIKI, 2010. Available at http://wiki.fot-net.eu/index.php?title=Smartway; http://www.nilim.go.jp/.

[SOC 03] SOCIETY I.C., IEEE Std 802.15.4-2003, Part 15.4: Wireless Medium Access Control (MAC) and Physical Layer (PHY) Specifications for Low-Rate Wireless Personal Area Networks (LR-WPANs), IEEE Computer Society, New York, NY, 2003.

[SOC 06] SOCIETY I.C., IEEE Std 802.15.4-2006, Part 15.4: Wireless Medium Access Control (MAC) and Physical Layer (PHY) Specifications for Low-Rate Wireless Personal Area Networks (WPANs), IEEE Computer Society, New York, NY, 2006.

[STO 99] STOJMENOVIC I., Home Agent Based Location Update and Destination Search Schemes in Ad Hoc Wireless Networks, SITE, University of Ottawa, 1999.

[STO 01] STOJMENOVIC I., LIN X., "Loop-free hybrid single-path/flooding routing algorithms with guaranteed delivery for wireless networks", *IEEE Transactions on Parallel and Distributed Systems*, vol. 12, no. 10, pp. 1023–1032, 2001.

[SUI 84] SUI H., ZEIDLER J.R., "Optimal transmission ranges for randomly distributed packet radio terminals", *IEEE Transactions on Communications*, vol. 32, no. 3, pp. 246–257, 1984.

[TOU 80] TOUSSAINT G.T., "The relative neighbourhood graph of a finite planar set", *Pattern Recognition*, vol. 12, no. 4, pp. 261–268, 1980.

[TSU 00] TSUGAWA S., "An introduction to demo 2000: the cooperative driving scenario", *IEEE Intelligent Systems*, vol. 15, pp. 78–79, 2000.

[TSU 05] TSUGAWA S., *Issues and Recent Trends in Vehicle Safety Communication Systems*, LATSS Research, 2005.

[WAN 05] WANG S.Y., *et al.*, "A practical routing protocol for vehicle-formed mobile ad hoc networks on the roads", *IEEE International Conference on Intelligent Transportation Systems*, Vienna, Austria, 2005.

[WAN 07] WANG W., XIE F., CHATTERJEE M., "An integrated study on mobility models and scalable routing protocols in VANETs", *Proceedings in 2007 Mobile Networking for Vehicular Environments*, Anchorage, Alaska, pp. 97–102, 2007.

[ZHA 06] ZHAO J., CAO G., "VADD: vehicle-assisted data delivery in vehicular ad hoc networks", *Proceedings IEEE INFOCOM 2006. 25TH IEEE International Conference on Computer Communications*, IEEE, 2006.

[ZOR 03] ZORZI M., RAO R.R., "Geographic random forwarding (GeRaF) for ad hoc and sensor networks: multihop performance", *IEEE Transactions on Mobile Computing*, vol. 2, no. 4, pp. 337–348, 2003.

[WG] 1609 WG – Dedicated Short Range Communication Working Group, 2013. Available at http://standards.ieee.org/develop/wg/1609_WG.html.

Chapter 3

CONVOY: A New Cluster-Based Routing Protocol for Vehicular Networks

3.1. Introduction

Intelligent transportation systems have been developed throughout the years mainly for the purpose of improving road safety. The so-called "intelligent" vehicles were thus equipped with sensors that assist the driving for the antiblocking system in case of braking (Anti-lock braking system (ABS)), route guiding, collision detection or the antiskidding mechanism (ESP – electronic stability program). Subsequently, vehicles were equipped with radio communication, which enabled them to communicate either with one another via (*vehicle-to-vehicle (V2V) communication*), or with the road side unit (RSU) installed along the road via *vehicle-to-infrastructure* (V2I)

Chapter written by Véronique VÈQUE, Florent KAISSER, Colette JOHNEN and Anthony BUSSON.

communication [PET 11]. In the field of road safety, communications help extend the detection range of embedded radar sensors. For example, a vehicle that detects the presence of black ice will be able to communicate this information to the vehicles that follow it. Having been warned, the drivers will thus be able to slow down earlier and therefore avoid skidding.

By generalizing intervehicular communications, we can constitute vehicular *ad hoc* networks (VANETs), which are a particular case of mobile *ad hoc* networks (MANETs). The VANETs thus facilitate the exchange of information between several vehicles, regarding the state of the road or the traffic in a given geographical area. However, it has been established that in the case of low vehicular density, a VANET has not always been a connected graph [SCH 06, WIS 07] and that it has been partitioned into several groups of vehicles disconnected from one another [MAB 06]. On the contrary, if the density becomes significant such as in a traffic jam, the network performance – in terms of offered traffic and delay – plummets. Finally, if the network extends upon several kilometers on the highway, the delay and the rate of packet loss *increase* with distance. These last two problems are strongly related to the issue of scalability, which we will discuss later on.

Besides intelligent transportation system (ITS) applications, user-oriented applications involving vehicular networks have also been developed. They include classic Internet applications (Web surfing, e-mail, computer games, Internet Protocol television (IPTV), etc.) as well as geolocalization applications for guidance or tourism. Using V2I communications, these applications need to access an operated infrastructure network, such as a long-distance mobile network, 3G or LTE (Long Term Evolution). However, the capacity of these mobile networks is often quite limited on the rural areas that traverse the highways. In addition, because the vehicles drive at high speed, the communications suffer from frequent handovers and run at a great risk of breaking up radio links. VANETs can thus be used to fix these problems by extending the coverage of operated mobile networks with an *ad hoc* part, we will therefore refer to as V2V2I for *Vehicle-to-Vehicle-to-Infrastructure*.

In this study, we are mainly interested in *V2V2I* communications, with a focus on organizing the *ad hoc* part of the network into groups of vehicles or *clusters*, each of them organized around a main cluster called a *clusterhead*. We aim to decrease the handover overhead for the mobile network. Indeed, we propose that the communications within the group be done in an *ad hoc* manner, whereas the communications between the groups should pass through the mobile network. The clusterhead therefore works as a gateway between its group and the mobile network. Being the only mobile that has access to a mobile network, the handover overhead reaching the base will be diminished. By using these principles, we have developed a self-organization protocol, which we will present in section 3.5.

The formation of groups is a problem of self-organization in *ad hoc* networks, known as *clustering or network partitioning*. Numerous works have been published on the subject, for *ad hoc* networks, sensors or MANETs [YU 05, AGA 09]. The main objective of the partitioning of an *ad hoc* network is to improve the protocol scalability. Thus, by dividing the space, the network becomes easier to manage and the coordination messages are no longer exchanged between all the nodes of the network, but only between the nodes of the group. If the nodes are mobile, they can change groups throughout time, which triggers the need for structural updates. Too frequent updates can make the performance of the hierarchical approach plummet, so it becomes a less interesting approach. We will therefore try to build groups that are as stable as possible, by choosing effective regrouping criteria. The partitioning approaches are well adapted to vehicular networks because the dynamics of road traffic triggers the formation of "natural" clusters at the intersections, or of convoys on the highway. For vehicular networks, as for ITS applications, the spatial or geographical dimension is an important factor. We show that the geographical position of vehicles is one of the important criteria in cluster-based protocols for vehicular networks. Since vehicles move in a space constrained by routes, other criteria are also significant such as speed and direction. Furthermore, in the context of radio communication, the quality of a connection or its stability is also important.

This chapter begins with an overview of the basic notions of partitioning in *ad hoc* networks as well as a presentation of the best well-known clustering algorithms. We will then review several proposals that are specific to vehicular networks. Then, we will present in detail CONVOY, our convoy formation protocol and its different algorithms. Finally, we will present an performance study of this protocol via simulations.

3.2. Clustering or network partitioning

3.2.1. *General remarks on the partitioning of mobile ad hoc networks*

The organization of an *ad hoc* network into node groups is known as clustering or network partitioning [YU 05]. The partitioning of the network organizes the network into node groups or clusters. Within the node group thus built, one particular node plays a special role: the *clusterhead*, or the leader of the group, organizes the communications within the cluster, serves as a gateway for communication with the other groups and connects them with one another.

Partitioning presents us with several advantages; the first one is the scalability of protocols. The division of the space indeed makes the network easier to manage and decreases the coordination messages exchanged between all the nodes. For example, the hierarchical routing uses a routing that is internal to the cluster, and a different type of routing between the clusters. The second advantage is the reduction of the execution charge of high level protocols on all the individual nodes. Finally, from the point of view of radio transmission, the partitioning while limiting the interactions between nodes also allows us to limit access collisions and interferences.

The first stage of partitioning is the clustering algorithm. The clustering algorithms can be classified depending on their way of controlling:

– the size of clusters defined either by the maximum number of nodes in the cluster, or by the size of the cluster in a geographical context (in kilometers);

– the maximum number of hops between each member and its clusterhead;

– the maximum number of clusters in the network;

– the mobility criterion of the nodes in highly dynamic networks.

The *clusterhead* is determined by a choosing algorithm, called *clusterhead election*, that can constitute the initialization of the cluster. Because the vehicle moves, periodic cluster reformations are needed and membership updates are performed using node adding functions or node retrieval functions.

In an *ad hoc* network, the partitioning approaches are naturally distributed and need numerous message exchanges for the formation and the maintenance of clusters. This cost of the partitioning is called "clustering overhead". The objective of a clustering algorithm will therefore be to limit this overhead as much as possible while ensuring the *stability* of the formed structure. Another desirable property that is researched in the case of node mobility is the capacity for quickly adapting to the structural changes that take place when the clusterhead or one of the nodes leaves the group, or when the radio connections are cut.

3.2.2. *Controlling the number of hops*

The distributed clustering algorithm (DCA) and distributed mobility-adaptive clustering (DMAC) algorithm [BAS 99] allow the formation of one-hop *clusters*, whose members neighbor at least one clusterhead. Lowest ID clustering was generalized to a weight-based clustering technique, referred to as DCA in [ALM 10]. In DCA, each node is assumed to have a unique weight (hence the weights are totally ordered) instead of just the node ID, and the clustering algorithm uses the weights instead of the IDs for the selection of clusterheads. However the technique of weight assignment has not been discussed. In DCA, each node is assumed to have a unique weight that can represent the degree of a node, the inverted value of its average speed, an identifier, etc. Hence, the *clusterhead* is defined as the heaviest weight node. This algorithm is well adapted to *ad hoc* networks whose topology is static or quasi-static. The DMAC algorithm is better adapted to dynamic networks, because it introduces an algorithm for the maintenance of the cluster, which reacts to the changes of medium access control (MAC) level topologies: new links and link failures. Moreover, it authorizes a larger number of clusters, thus ensuring a greater stability of the hierarchical structure. With a one-hop cluster, the intracluster communications do not need any routing and are extremely simplified. However, these algorithms can lead to a large number of clusters. Finally, a one-hop cluster will have limited coverage, which will trigger frequent updates of the cluster in case of mobility. To minimize the number of clusters, the authors of [NOC 03] have generalized the one-hop algorithm to k-hops clustering. To discover the proximity to k-hops, the cluster update messages are broadcast with a maximum number of hops or Time To Live delay (TTL) equal to k.

3.2.3. *Controlling the number of nodes*

It is also possible to limit the size of a cluster by defining the maximum number of nodes in a cluster. Indeed, maintaining clusters of a balanced size facilitates the balance of communications between the nodes and reduces the collisions within the cluster. The algorithm presented in [OHT 03] limits the size of the cluster. If the cluster is too small, a node from the outskirts of a neighboring larger node is then recruited.

The formation of clusters in an *ad hoc* network is not an end in itself but a means of improving the *ad hoc* routing by limiting its research time or its signaling overhead. Thus, Niu *et al.* [NIU 06] propose an *ad hoc* routing protocol based on a cluster formation algorithm with k-hops. The *ad hoc* network is hierarchized upon two levels: the higher level (the level of the clusters) and the lower level (the level of the nodes). At the cluster level, the routing is proactive: each clusterhead keeps a graph of the set of clusterheads of network clusterheads and a list of the members of the associated cluster. Another strategy for improving the performance of a reactive routing is proposed in [OHT 07] with the formation of clusters according to [OHT 03]. A clusterhead builds a minimum spanning tree based on the topology information of its cluster, information obtained from the messages sent periodically by the members of its cluster. The messages are disseminated throughout the tree and not by inundation, thus reducing the overhead.

3.2.4. *Role of the clusterhead*

The partitioning of the network has two levels of communication: intracluster communication and intercluster communication. The intercluster communications are often long-distance and federate the clusters among themselves. Most often, they are carried out due to a single node per cluster called a clusterhead. We usually consider the following communication scenario: the messages for a node of the same cluster are broadcast in the cluster whereas those for a faraway node are transmitted to the clusterhead that will further transmit them to the target cluster. The choice of the clusterhead within a cluster can be made due to certain characteristics (i.e. double radio interface and throughout), which justifies this gateway role, or by a distributed election algorithm between the nodes that discerns between these characteristics based on one of such criteria: identifier, remaining power, the shortest distance, etc. In the state of art that we will soon present, we specify the way in which the clusterhead is chosen according to the regrouping criterion. We will now specify the differences between a VANET and a fixed or little mobile *ad hoc* network.

3.3. Mobility-based clustering in *ad hoc* vehicular networks

As we have already mentioned, clustering approaches are well adapted to vehicular networks because of the formation of "natural" groups at the intersections or of convoys on the highway. The difficulty in clustering on vehicular networks is to ensure the stability of the constituted groups. If we build a group according to the single criterion of geographical localization, the difference of speed between the vehicles runs the risk of immediately triggering an update, with adding and retrieving members in the group. Too frequent updates increase the clustering overhead and also diminish its interest. This is why a lot of hierarchical protocols for VANET not only consider spatial information such as the position or the quality of radio communication, but also temporal information regarding the behavior of the vehicles, such as the direction of their movement and their speed. The result is a lower updating rate, that is a property that certain works refer to as the "resilience" property of the network [SOU 10]. These clustering protocols allow us to introduce stable structures in a very unstable environment.

In the following section we present the main spatial and temporal regrouping criteria as well as several routing protocols called "mobility-based clustering", which have used or combined these criteria in order to build stable clusters.

3.3.1. *The dynamics of vehicular traffic in VANETs*

In a highly dynamic network such as VANET, the mobility of the nodes jumbles up the hierarchical structure and triggers frequent reorganizations of the groups [GHO 08]. This instability results in a larger number of exchanged messages and in poor performances. The stability property of the clusters is of primary importance for these dynamic networks. In [LEE 10], the comparison of the routing protocols for the VANETs shows the superiority of the *geographical* or *position-based* protocols. To maintain the stability of clusters, the position of the vehicles is not enough and many propositions use both spatial and temporal information, which allows us to represent the particular dynamic of the moving vehicles and to build clusters that have a longer lifetime. The main objective of the protocols called "clustering mobility-based" protocols is then to group together the vehicles that have the same dynamics, whether they are spatial or temporal. The spatial dependencies are represented by the position of the vehicle and its distance from another point, its moving direction or the quality of its radio link with another vehicle. The temporal dependencies are the moving speed of the vehicle and, above all, the relative speed between two vehicles.

For several years now, vehicles have often been equipped with global positioning system (GPS) receivers, which give the latitude and longitude coordinates of the vehicle. However, the precision of this mainstream equipment is only 5–10 m. The technologies integrated in intelligent vehicles now allow us to improve the positioning accuracy especially due to map positioning, and the sensors embedded in the vehicle such as distance radars and the odometer [CAP 06, DEA 09]. Different propositions of clusters based on the position or mobility use these precise localization techniques.

We also assume that the neighbor discovery procedure is always available in considered vehicular *ad hoc* protocols. Thus, each node is able to inform its neighbors while periodically disseminating HELLO messages throughout its radio coverage area, and therefore is able to know all of its neighbors. These HELLO messages contain at least the node's identifier, but also state information, such as its position and the list of its neighbors. The position of the vehicle is therefore known due to its embedded GPS device but this position cannot, by itself, act as a regrouping criterion because vehicles change their position very quickly and therefore, change their cluster just as quickly. In the following sections, we present clustering algorithms based on spatial or temporal criteria that have been proposed for improving the routing protocols of vehicular networks.

3.3.2. *Clustering according to the lane*

Almalag and Weigle [ALM 10] refine the position of the vehicle with the exact knowledge of the lane that the vehicle drives on. The lane is obtained by checking the position of the vehicle against a precise map of the streets. A cluster is built for each lane depending on its future direction after an intersection "turn left, turn right or straight ahead"; this is also helpful for emergency brake applications.

3.3.3. *Clustering depending on the relative speed between the vehicles*

Another regrouping criterion that results from the position of the vehicles is the relative distance between the vehicles or their pertaining to a given geographical area [BON 07, JAY 12]. The objective is to limit the message dissemination in a constrained space, whether it be for the routing or for the ITS applications whose coverage range is local. The Euclidean distance between two nodes is calculated as the difference between their respective coordinates. However, there is an oscillatory effect upon the distance between the vehicles. In [BLU 03], this effect is taken into account with a maximal (geographical) distance criterion between two clusterheads and in [JAY 12], the authors also add a criterion based on the point where the vehicle exits the road.

3.3.4. *Clustering depending on the direction of the movement (movement-based)*

It has been noted very early that vehicles moving in opposite directions could not communicate for too long [DUC 07, MEN 07]. Even worse, when regrouped in the same cluster, these vehicles stay for a very short period in the cluster and make the structure unstable. The direction is obtained from the history of the vehicle's positions, with or without a map, or due to the motion angle compared to the north. Several recent works thus use the direction as a criterion for regrouping vehicles.

In [TAL 06], the vehicle-heading-based routing protocol (VHRP) creates four groups of vehicles according to their direction, with the aim of building stable routes that preferably cross between the vehicles of the same groups. The clusterhead is the head vehicle. Inside the group, a road is established between all of the vehicles within the same group. Simulation shows that by varying the speed, VHRP reduces packet loss in comparison with classic reactive protocols. However, the analysis has not been carried out sufficiently in depth. Fan *et al.* [FAN 07] extend the work of [FAN 05] for considering the direction of the vehicles' movement: two vehicles going in opposite directions cannot be in the same *cluster*. The election of the clusterhead is based on the previous situation of the nodes, old clusterheads being preferred over new clusterheads.

3.3.5. *Clustering depending on the radio link quality*

In [MEN 07], the movement prediction-based routing (MOPR) protocol predicts future positions of the vehicles for each routing pathway between a source and a destination relying on their positions, their speeds and their directions. Furthermore, MOPR is capable of estimating the lifetime of radio connections in the network and then chooses the most stable route between the source and the destination, a route that is made of the most stable links. MOPR is seen as a function that was integrated in different *ad hoc* routing protocols, such as *ad hoc* on-demand distance vector (AODV) or greedy perimeter stateless routing (GPSR). This approach was revisited and applied to the constitution of clusters with a longer lifetime in [JAY 12]. This algorithm uses not only the lifetime of the connections, but also the position, speed and direction criteria for calculating a score that allows us to select the best hop for forwarding a packet. In the beginning, all of the nodes are in the same cluster, and the cluster is then divided into smaller and smaller groups, until each group meets the distance condition (250 m) and the vehicle number condition. Within each minicluster, the score is used for choosing the node that will forward the packet.

3.3.6. *Clustering depending on speed and relative speed*

The vehicle speed was also taken into account in order to regroup the vehicles in convoy because a convoy of vehicles with quasi-identical speed is a relatively stable structure. The mobility criterion is more precise and is called, depending on the work, *aggregate mobility* or *relative speed*. Along with direction, it is the criterion that has most improved the stability of clusters. Thus, Basu *et al.* [BAS 01] have introduced the concept of aggregate mobility in order to improve the stability of clusters in their MOBIC algorithm. The authors propose a relative mobility metric system for MANETs. Relative mobility is measured using the received signal strength (RSS), which is an indication of the distance between the sender and the receiver. The strength received is exchanged between the nodes by means of the neighbor discovery procedure. The relation between two successive measures provides the relative mobility metric. The election of the clusterhead is made according to the *lowest ID* algorithm where the identifier is the variation of the relative mobility. The changing of the clusterhead is realized if two clusterheads are within the range of each other and the two clusters are merged together. A member that is out of the range of its clusterhead leaves the cluster and can join another cluster. Other propositions have improved this basic idea. Thus, in [FAN 05], the authors take into account the speed or the position of the neighboring vehicles in the election algorithm of the clusterhead (lowest ID and highest degree k-hops), in order to increase the stability of the cluster.

3.3.7. *Clustering depending on the position, speed and direction*

Shea *et al.* [SHE 09] regroup the vehicles with the help of an affinity function that considers the position of the node, its direction and its speed, all at the same time. For different update intervals of the cluster and different speeds, [SHE 09] displays stability metrics such as the average lifetime of the clusterhead, the average lifetime of the members, the clusterhead change rate and the average number of clusters compared to the MOBIC protocol [BAS 01]. The results show that their approach indeed substantially improves cluster stability.

Souza *et al.* [SOU 10] uses the same principle but proposes a more precise mobility metric called aggregate local mobility (ALM) always with the hypothesis that every vehicle knows its geographical position. Similarly, the vehicles include their position in the HELLO messages disseminated in their neighborhood. The ratio between two successive distances defines the relative mobility between two vehicles. ALM forms coherent vehicle clusters of the same speed and the same direction whose performance analysis demonstrates their stability. In [RAW 09], the authors use the same regrouping criteria and develop a new multicriteria leader election algorithm, then show the validity of the approach via modeling.

As we have just seen, when we have a lot of information on the movement and the position of vehicles, it is possible to conceive the hierarchical protocols that are stable and that minimize the overhead of the cluster updates. However, taking into account multiple criteria leads to complex algorithms. This aspect is never mentioned in the references that we cite. In vehicle networks, the partitioning of the network is not only used for improving routing but it is also used at other levels of the protocol architecture. Thus, we find clustering techniques at MAC level and at the application level for the localization or for other applications in the field of transport. We describe these propositions in the following section.

3.4. Clustering of VANETs for MAC and transport applications

3.4.1. *Cluster-based MAC protocol*

MAC hierarchic protocols seek to alleviate several problems induced by the strong mobility of vehicles. By regrouping the vehicles of comparable speed in clusters, we can reduce the relative speed (as defined in section 3.3.6) of communicating nodes. They bring a certain improvement for a dense network of vehicles, for example during a traffic jam. The limited number of nodes that are competing in the cluster reduces the contentions and thus the congestion of the network, and minimizes the influence of the hidden terminal problem. The Quality-of-Service can, moreover, be guaranteed inside the cluster.

The cluster-based medium access control (CBMAC) protocol [GUN 07] was one of the first hierarchical MAC protocols to be proposed. The members of the cluster come inside the cluster and leave the cluster depending on a weight, which in turn depends on their speed, their distance from the clusterhead or their connectivity. The role of the clusterhead is then to allocate the shared bandwidth using the time division multiple access (TDMA) scheme to the different members of the cluster. [RAW 08] and [SHE 09] are two other examples of the use of clustering that improves the MAC protocol.

3.4.2. *Clustering for transport applications*

Applications for the use in transport (ITS) often have geographically limited coverage, whether with regard to road safety or the state of road traffic. The coordinates of the sender (latitude and longitude) are thus transferred at the same time as this information. We have seen that it was easy to know the position of the vehicle itself by means of a geolocation device (GPS). On the contrary, it is necessary to use a location service in order to know the position of a faraway destination. This research is costly because it is most often done in a greedy manner [DAS 05]. The cluster formation techniques can improve the scaling of a geographical location protocol in a vehicular network. In [SAN 06], the authors propose a two-hop cluster formation algorithm with clusterhead election, which can serve as a gateway. The overhead of the localization is diminished because only the gateway nodes transmit the localization requests.

Another family of transport applications uses the dissemination of traffic data or security data in a limited area. For example, the *floating car data (FCD)* applications [LIN 06] consist of collecting the different information produced by the vehicle sensors (position, speed and direction) and of disseminating them to neighboring vehicles in such a way as to enrich global knowledge of the road scene. Among the hierarchical approaches developed for data dissemination, we have chosen to present: [MAS 11b],

[OHT 12] and [REM 12]. In [MAS 11a], the authors consider a dissemination application that seeks to dynamically control the signaling lights at the intersections. This dissemination relies on two hierarchical protocols called Clustering in Direction in Vehicular Environment (C-DRIVE) and Modified C-DRIVE (MC-DRIVE) [MAS 11b], whose clusters are made of the vehicles close to the intersection and depending on their future direction after the intersection (left, right, straight ahead and U-turn). The size of the cluster is limited to a certain distance from the intersection. At this distance, each vehicle disseminates a HELLO message containing its identifier, its direction and its driving lanes provided that it seeks another vehicle in front of it. The two protocols differ in their way of choosing the clusterhead.

In [OHT 12], the authors propose a cluster-based data transfer scheme, using the position and the direction of the movement of vehicles. Inside a cluster, the data packets are disseminated epidemically starting from the clusterhead and ending with the destination vehicle. When two clusters are within range of each other, each cluster decides if it must or must not transfer its packets depending on their position, their direction and the position of the destination. The construction and the maintenance of clusters are not detailed by the authors.

In [REM 12], the authors also regard the FCD-type dissemination applications for the fourth-generation mobile network standard, LTE Advanced. Taking advantage of the fact that the LTE standard intrinsically regards the communication between the terminals (and thus of the signaling), the LTE4V2X protocol forms the groups in a centralized manner, via the base station (or e-NodeB) [REM 11]. The vehicles are supposed to dispose of an LTE interface and an IEEE 802.11p interface at the same time. The double objective is to decrease the load of the traffic going upward toward the base station and to extend the network coverage due to the local communications between vehicles. The formation of the cluster takes into account two criteria: speed and direction. The maintenance and the update of the groups is periodic, which makes the overhead insensitive to the number of vehicles or to group changing. On the contrary, the supplementary computational load on the eNode-B has not yet been studied. The stability property of the clusters or their lifetime have not been highlighted either, while each eNode-B must regularly update the clusters.

The different works that we have presented highlight the need to take into account the spatial and temporal characteristics of the vehicles for forming stable clusters. In [MAB 06] and [WIS 07], the road traffic analysis carried out on the highway shows that the vehicles have a tendency to regroup naturally in a convoy in the same direction and with the same speed. In the following sections, we propose an approach called CONVOY that is based on these natural convoys and organizes them starting from criteria such as position, relative speed, movement direction and the quality of the radio connection.

3.5. CONVOY: a vehicle convoy formation protocol

On the highway, the vehicles follow each other in a line. In [MAB 06], we have shown that the particular dynamic of vehicles in such an *ad hoc* network triggered the regrouping of the vehicles in clusters that are disconnected from one another. The convoy formation is thus natural in a highway context. In a convoy where the vehicles follow each other, the clusterhead is thus thoroughly designated: it is the head of the convoy. We consider that a vehicle is equipped with a single wireless interface, a highly performing geolocalization system using a GPS terminal as well as information specific to the vehicle such as its position, its actual speed and its direction.

CONVOY combines more criteria, as presented in section 3.3. To ensure the stability of the convoy, the membership of a new vehicle to the convoy depends on:

– its position;

– its relative speed;

– its movement direction;

– the duration of its radio connection with the other concerned members of the convoy.

Inside the convoy, the communications can be made in several hops. Our algorithm therefore forms k-hops clusters. To limit the congestion inside the cluster, we have also defined a maximum convoy length in kilometers without limiting the number of vehicles that are present in the convoy. Inside the cluster, the graph representing the network must then be connected and the intervehicle communication will be ensured by means of an intracluster routing protocol. We will not consider the lane on which the vehicle is driving, the precision of the geolocalization systems being to a few meters. However, the regrouping of vehicles on the same lane is helped by the criterion for joining a convoy, as we will discuss in the following.

Figure 3.1 illustrates these principles: the vehicles that drive in opposite directions are not in the same convoy. However, in convoy 7, the vehicles are in the same convoy while they are on different lanes. The speed of the vehicles in convoys 1 and 4 is very different, which explains the fact that they are in different convoys.

With the aim of simplifying, we have chosen to separate the convoy formation protocol from the intervehicle communication protocol.

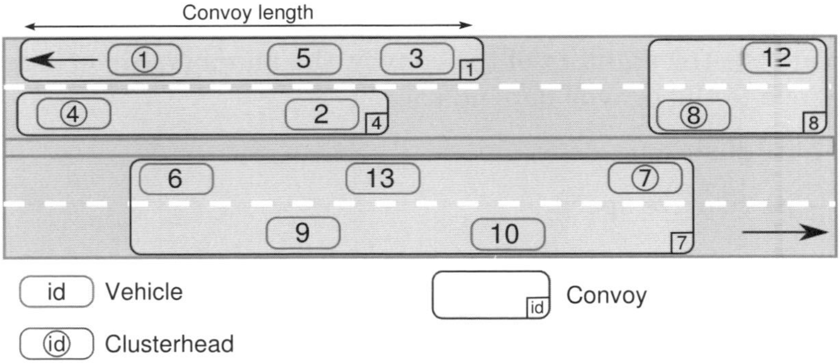

Figure 3.1. *Vehicle convoys on a 2 × 2 way highway. The head of the convoy is the vehicle in front of all the other vehicles*

3.5.1. *Intra-convoy communication protocol*

Inside the convoy, the communication protocol is the simplest possible and draws inspiration from [DUC 07]. It is based on the geographical position of the vehicles and the direction of the convoy. The dissemination of a packet in the direction (respectively in the opposite direction) of the movement of the convoy allows us to deliver a message to all the vehicles in the back of the convoy (and respectively in the front of the convoy) of the vehicle that initiates the message. The CONVOY protocol therefore is not in charge of sending messages.

3.5.2. *Convoy formation algorithm*

Our algorithm for forming convoys revisits the principles of the cluster formation algorithms. The head of the convoy, or the clusterhead, is simply the vehicle at the *head of the convoy*, that is the vehicle present in front of all the vehicles members of the convoy (Figure 3.1). Contrary to the algorithms described previously, which limit the size of the cluster to a certain number of hops or to a certain number of nodes, our algorithm defines the maximum size of the cluster via its geographical length (in meters), that is the maximum distance separating the head of the convoy from the queue of the convoy (Figure 3.1).

Initially, every vehicle is alone in its own convoy and thus at the head of the convoy. Therefore, there are as many convoys as there are vehicles in the network. A node states its presence to its neighbors by sending a HELLO message at regular intervals, containing its identifier, the identifier of its convoy, its position, its direction and its speed.

The formation and the maintenance of the convoys are ensured by two distinct procedures: the *merger* and the *division* of convoys. Merging allows a convoy to grow by recruiting new vehicles. The division is triggered when the convoy is no longer connected: there is a node that can no longer communicate with the head of the convoy. The set of messages used by these procedures is shown in Table 3.1.

Name of the message	Acronym	Dissemination mode
Join convoy request	JOINREQ	Movement direction
Join convoy reply	JOINREP	Towards a destination
Node join	NODEJOIN	Towards a destination
Convoy info	CONVOYINFO	Direction opposed to movement
Carrying out a CI	INFOACK	Movement direction
Hello	HELLO	Vicinity

Table 3.1. *List of messages used by the convoy formation algorithms*

3.5.2.1. *Convoy merger*

In the following, we describe the procedures triggered upon receipt of a HELLO message. Let us remember that in an *ad hoc* network, the HELLO messages are disseminated periodically through a node to its neighbors in order to inform it of its presence and to exchange information regarding the state that will be exploited by the protocols.

Algorithm 3.1 Variables, constant and initialization of a node

---

Constants :

Id ▷ Vehicle identifier

$LifeTimeLimit$ ▷ Minimal lifetime of a connection

$ConvoyLengthLimit$ ▷ Length limit of a convoy

Node variables :

$IdConvoy$ ▷ Identifier of the convoy to which we belong

$ConvoyHead$ ▷ True if the vehicle is the head of the convoy

$ConvoyLength$ ▷ Actual length of the convoy

$JCReqTable$ ▷ Set of sent JOINREQ messages

$ClusterNodes$ ▷ Set of vehicles belonging to the cluster

External variables :

$CurrentSpeed$ ▷ Current speed vector

$CurrentPos$ ▷ Current position

procedure INITIALIZATION

$IdConvoy \leftarrow Id$

$ConvoyLength \leftarrow 0$

$ConvoyHead \leftarrow True$

$JCReqTable \leftarrow \emptyset$

$ClusterNodes \leftarrow \{Id\}$

end procedure

To initiate the merging of two convoys, C_i and C_j, the *clusterhead* of C_i must be the *neighbor* (*clusterhead* or not) of a vehicle contained in the *cluster* C_j. Starting from the information contained in the HELLO messages received from its neighbor, the *clusterhead* decides to become a candidate for the merger or not (algorithm 3.2).

Algorithm 3.2 Reception of a Hello message. Parameters: identifier, position and speed of the node. Identifier of the convoy

```
function LIFETIME(speed, pos)
    d ← |pos − CurrentPos|
    if speed < ||CurrentSpeed|| then
        return (R − d)/(||CurrentSpeed|| − speed)
    end if
    if speed > ||CurrentSpeed|| then
        return (R + d)/(speed − ||CurrentSpeed||)
    end if
    return ∞;
end function

function SAMEDIRECTION(v1, v2)
    return v1 × v2 > 0            ▷ Test if the scalar product is positive
end function

procedure HELLO(id, idConvoy, pos, speed)
    if ConvoyHead
        ∧ LIFETIME(||hello.speed||, hello.pos) < LifeTimeLimit
        ∧ SAMEDIRECTION(hello.speed, CurrentSpeed)
        ∧ IdConvoy ≠ hello.idConvoy
        ∧ CurrentPos < hello.pos
        ∧ hello.idConvoy ⊈ JCReqTable then
         SEND JCREQ(hello.idConvoy, ConvoyLength)
    end if
end procedure
```

Following are the admission conditions:

– The lifetime of the connection between the clusterhead and its neighbor must be higher than the limit *LifeTimeLimit* fixed by the protocol, which allows us to guarantee a lifetime of the cluster that is superior to that limit (supposing that the speed of the vehicles is constant).

– The two convoys C_i and C_j move in the same direction.

– The neighboring vehicle is in front of the *clusterhead* of C_i.

– A merging candidacy between C_i and C_j did not already take place.

After the merger, the clusterhead being always in front of all the vehicles' members of the convoy, the clusterhead of C_i becomes a member of the convoy C_j.

If all of these conditions are met, the proposed clusterhead C_i sends a JOINREQ message to the neighboring vehicle that sends it to its clusterhead, called the *neighboring clusterhead.* The neighboring clusterhead verifies the size of the two joined convoys (algorithm 3.3): if it is lower than the maximum size of the convoys, then it sends a response message JOINREP to the clusterhead candidate confirming the merger. This message is then sent toward the geographical position of the destination. If the size is higher than the maximum size, then it does not do anything and the merging process is aborted. Upon receipt of the message JOINREQ (algorithm 3.5), C_j sends a NODEJOIN message to the neighboring clusterhead C_i containing the list with the members of its convoy. Next, C_j disseminates a CONVOYINFO message in the direction opposed to the movement of the convoy and updates the identifier of the convoy to which it belongs.

Upon receipt of the NODEJOIN message (algorithm 3.4), the clusterhead adds the vehicles contained in the message from its list. Upon receipt of a message CONVOYINFO (algorithme 3.7), each vehicle updates the identifier of its convoy.

Algorithm 3.3 Receipt of a JOINREQ message. Parameters: identifier of the convoy, size of the convoy (boundaries)

```
procedure JCREQ(idConvoy, ConvoyLength)
    if ConvoyHead
        ∧ jcreq.ConvoyLength ∪ ConvoyLength < ConvoyLengthLimit then
        SEND JOINREP(jcreq.idConvoy)
    end if
end procedure
```

Algorithm 3.4 Receipt of a NODEJOIN message. Parameters: set of nodes of the convoy to be joined

```
procedure NJ(nodes)
    ClusterNodes ← ClusterNodes ∪ nj.nodes
end procedure
```

Algorithm 3.5 Receipt of a JOINREP message. Parameters: identifier of the convoy

```
procedure JCREP(idConvoy)
    IdConvoy ← jcrep.idConvoy
    ConvoyHead ← False
    SEND CONVOYINFO(IdConvoy)
    SEND NODEJOIN(ClusterNodes)
end procedure
```

3.5.2.2. *Division of the convoys*

The management of the division of convoys allows us to maintain the convoy. The detection of the link failure with a neighboring member is not enough for deducing that the vehicle leaves the convoy. Indeed, it can still communicate with all the members of the convoy and the clusterhead via the intermediary of another link. By definition, the division of a convoy takes place when a vehicle member can no longer communicate with the clusterhead, that is when there is no lane toward the clusterhead (Figure 3.2). The network being linear, none of these vehicles behind the disconnected vehicle can communicate with the clusterhead any longer. The head member thus becomes clusterhead and the set of the vehicles behind this vehicle rejoins the new convoy.

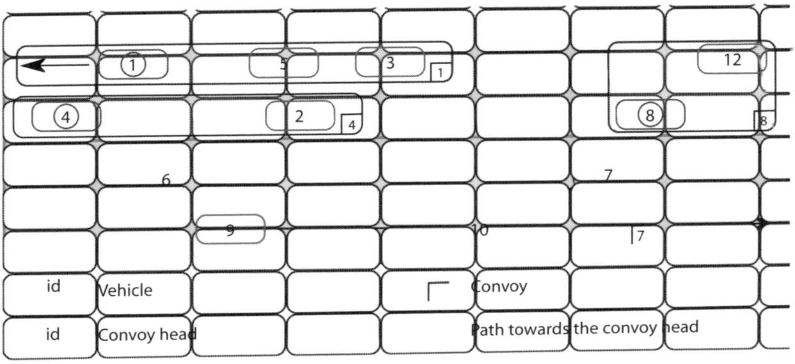

Figure 3.2. *Division of convoy 7. After the division, convoy 7 leaves and a new convoy 9 is formed.*

The *clusterhead* disseminates a message CONVOYINFO to all members of the convoy, within regular intervals, in the direction opposite to the movement of the convoy. When a member does not receive this message during a predetermined delay, it considers that there is no longer a lane with the clusterhead; the convoy being no longer connected, there is a division of the convoy. The member designates itself chief *clusterhead* and disseminates a CONVOYINFO message in the direction opposite to the movement of the convoy, allowing vehicles behind it to change the convoy (algorithm 3.6).

Upon receipt of a CONVOYINFO message (algorithm 3.7), a member allocates its convoy identifier to the new identifier of the clusterhead and acknowledges the reception of the message to its new clusterhead. Upon receipt of the acknowledgment, the new clusterhead adds the new member to its member list. The clusterhead in the back, not having received any acknowledgment of the CONVOYINFO message, considers that its members are no longer part of its cluster and removes them from its member list. The division procedure is thus terminated.

Algorithm 3.6 Upon convoy breaking

procedure CONVOYBREAK($idConvoy$)
  $IdConvoy \leftarrow Id$
  $ConvoyHead \leftarrow True$
  SEND CONVOYINFO($idConvoy$)
end procedure

Algorithm 3.7 Receipt of a Convoy info messages. Parameters: Convoy identifier

procedure CI($idConvoy$)
  $IdConvoy \leftarrow ci.idConvoy$
  $ConvoyHead \leftarrow False$
  SEND INFOACK
end procedure

3.6. Assessment of the convoy formation protocol

We have implemented the CONVOY protocol in the JIST/SWANS simulator [BAR 04]. To finely simulate the movement of the vehicles on a highway, we use the road traffic simulator developed in [MAB 07] to model the mobility of vehicles in a highway context. Besides the parameters connected to the road traffic (highway length and vehicle density), we use four parameters connected to the simulation of convoy formation whose default values are given in Table 3.2. All of the simulations were realized with a confidence interval of 95%.

Simulation time	600 s
Hello interval	2 s
Lifetime limit of a convoy connection	600 s
Convoy length limit	2 km
Highway length	10 km
Vehicle density	10 vehicles/way/km
Radio coverage	250 m
Bandwidth	11 Mb/s

Table 3.2. *Default simulation parameters*

We are interested in five metrics: the number of convoy breaks, the number of convoys, the partitioning rate, the length and the size of a convoy. The *number of convoy breaks* during a simulation allows us to evaluate the global *stability* of the convoys formed: the more significant the number of breaks, the less stable the convoys. The number of convoys allows us to evaluate the quality of their formation: in the worst-case scenario, there are as many convoys as there are vehicles. Each convoy thus being composed of a single vehicle, the formation of a convoy therefore does not present any interest, and the regrouping criteria are not valid. The smaller the number of convoys, the more significant the convoys and hence the more the mobility of the vehicles can trigger connection failures and convoy divisions. We are also interested in the partitioning rate defined as the number of convoys divided by the number of nodes. The partitioning rate is thus the reverse of the average size of convoys. We differentiate between *convoy length* and *convoy size.* The length of a convoy is the distance in kilometers between its head and its rear (i.e. the vehicle behind all the members of the convoy). The size of a convoy is the number of its member vehicles.

In the first stage, we seek the optimal parameters of the CONVOY. Then, we analyze the stability and the quality of the convoys formed with the optimal parameters we found.

3.6.1. *Optimal parameters of the algorithm*

We seek to regulate two significant parameters in CONVOY, namely the maximum convoy length in meters and the minimal lifetime of a connection for a node to be accepted in a convoy.

3.6.1.1. *Maximum length of a convoy*

The connectivity of a convoy is directly proportional to its density. The higher the density of a convoy, the larger the size of a convoy (potentially), thus decreasing the partitioning rate. This hypothesis is verified against the results in Figure 3.3(a): the partitioning rate decreases along with the density. In this same figure, we see that the partitioning rate is equally proportional to the length limit of the convoy. Consequently, the maximum length parameter of the convoys allows us to limit the size of the convoys. The variation of the partitioning rate is weaker as soon as the length limit is higher than 2,000 m and quasi-constant starting from 3,000 m. The number of convoy breaks increases with the density and the length limit of the convoys (Figure 3.3(b)). For a density of 24 vehicles/km/lane, the number of breaks explodes starting from 2,000 m. Beyond 3,000 m and with a density of 24 vehicles/km/lane, the strength required by the simulator is too significant to be able to carry out a simulation, hence the absence of the measures. We conclude that it is not useful to fix the maximum length of a convoy beyond 2,000 m in order to satisfy the compromise of a partitioning rate close to a minimum, and of a minimum number of convoy breaks. We have therefore set the limit of the convoy length to a default 2,000 m.

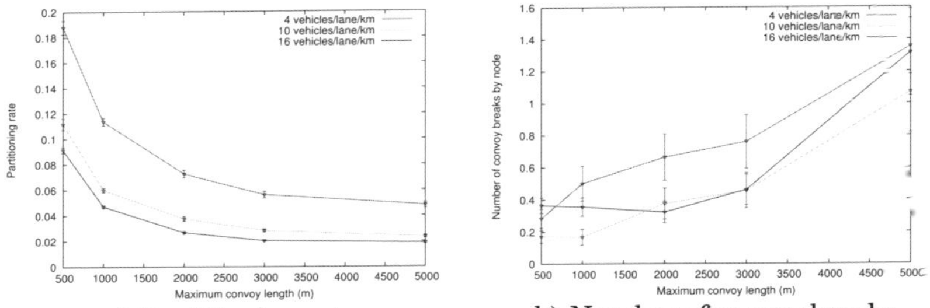

a) Partitioning rate b) Number of convoy breaks

Figure 3.3. *Performance depending on the maximum convoy length for several densities*

3.6.2. *Distribution of the length of convoys*

Having discussed the maximum length of the convoy, we must verify that it is valid. The distribution of the length of convoys is calculated at the end of the simulation. Figure 3.4 presents this distribution for a limit fixed at 2,000, 3,000 and 5,000 m. Each bar represents the number of convoys comprised between a length l and $l + 500$ m on a sample of approximately 100 convoys carried out in several simulations.

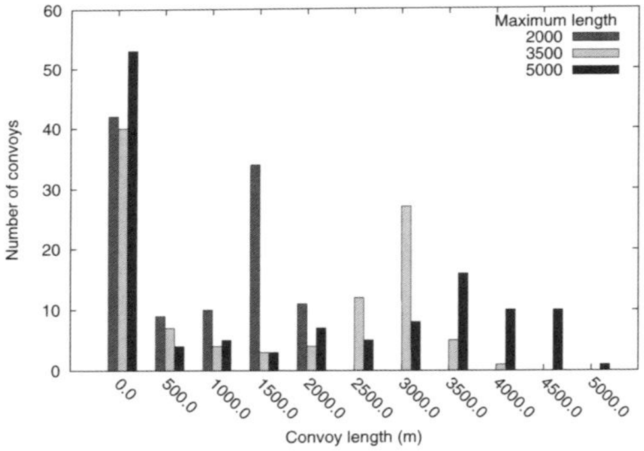

Figure 3.4. *Convoy length distribution in meters*

Let us note two glaring anomalies: a significant proportion of convoys of length comprised between 0 and 500 m, and convoys that are of a larger size than the set limit. When a node arrives on the highway, it is the head of the convoy and alone in the convoy (no member has yet joined it). It remains alone for a limited amount of time before merging with another convoy. Each node having been at the start in a zero length convoy, the proportion of small convoys close to the value zero in length (one single node in the convoy) is obviously important, but not significant. Indeed a convoy stays for very little time with a size less than 10, as we may observe in Figure 3.4. The second perceived anomaly is the existence of convoys of a size greater than the size limit (2,000 m). The speed of the vehicles being quite close to one another, we set the length limit of the convoy solely at the time of its formation, that is during merger. On the other hand, the topology of the convoy is dynamic: the relative distance between the vehicles varies with time. It is possible that a vehicle with a lower speed moves away from the other vehicles in the convoy and increases the length of the convoy over time. Consequently, although no new vehicle has joined the convoy, the length of the convoy can increase without causing a convoy break because the length remains close to 2 km. The distribution is in accordance with the waiting times, more so for the convoys of 2,000 and 3,000 m in length, the maximum being, respectively, for the convoys of a length comprised between 1,500 and 2,000 m and 3,000 and 3,500 m. For a limit of 5,000 m, the distribution is more widespread, maintaining the convoy higher than 3,000 m being more difficult because of the higher probability of a convoy break (Figure 3.3(b)).

3.6.3. *Convoy stability*

Our objective is to minimize the number of convoys thus formed, which involves large-sized convoys. On the other hand, the number of convoy breaks depends on the size of the convoys (Figure 3.5(b)). The connectivity of a convoy is directly proportional to its density. The denser a convoy, the larger its potential size, thus decreasing the partitioning rate. This hypothesis is verified by the results in Figure 3.3(c): the partitioning rate decreases with the density.

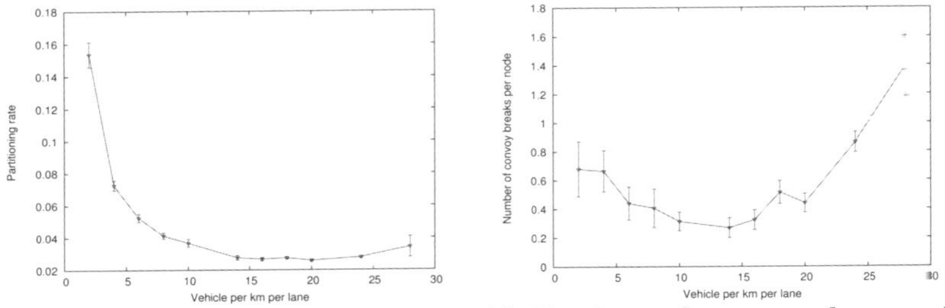

a) Density versus rate

b) Number of convoy changes by node

Figure 3.5. *Performance per node depending on the density*

In this same figure, we see that the partitioning rate is also proportional to the length limit of the convoy (Figure 3.3(a)). Consequently, the maximum length parameter of the convoys allows us to limit the size of convoys. The variation of the partitioning rate is lower since the length limit is higher than 2,000 m and quasi-constant starting from 3,000 m. We have found a compromise with a convoy limited at 2,000 m.

We are therefore interested in the length of a convoy with a size n, and a length limit of 2,000 m. In Figure 3.6, we measure the distribution of the lifetime of a convoy depending on its size. Here, the lifetime is expressed in a simulation percentage, namely the period of time during which the convoy has a size comprised in the interval. The repartition of the size of the convoy depends on its density:

– for 4 vehicles per kilometer and per lane, a cluster comprises up to 40 vehicles;

– for 10 vehicles per kilometer and per lane, a cluster comprises up to 80 vehicles.

– for 16 vehicles per kilometer and per lane, a cluster comprises up to 140 vehicles.

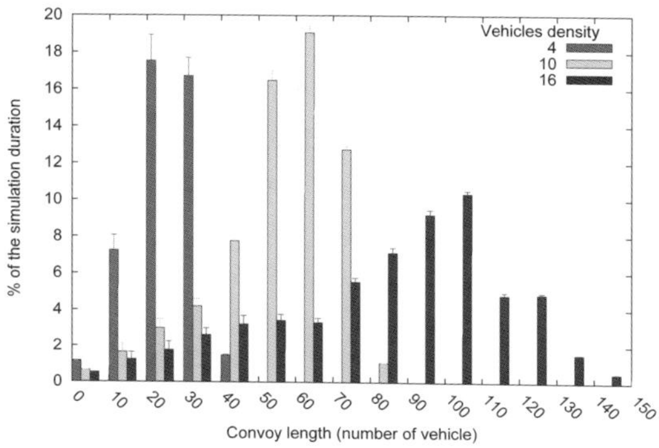

Figure 3.6. *Distribution of the lifetime of the convoys depending on their size*

Let us also note that the large size convoys are those that last the longest, especially with the densities of 4 and 10 vehicles per kilometer and per lane. This property is very helpful for obtaining stable convoys of a size sufficiently large in order to justify convoy formation in a network of vehicles.

3.7. Conclusion

The organization in clusters is well adapted to VANETs because natural groups form at the intersections or on the highway. However, VANETs are very dynamic networks and the partitioning of the network must be made according to representative criteria of the spatial and temporal behavior of the vehicles in motion. The objective is to introduce stable structures in a globally unstable system. Among the criteria used in the different protocols and clustering algorithms that we have presented, it is worth highlighting: the geographical position, the distance between vehicles, the speed and the average speed within the group and the direction of the movement. We have then presented propositions in each of the different frameworks of use such as the MAC level or the ITS applications.

Finally, we have presented CONVOY, an original convoy formation protocol adapted to the vehicular network. The formation of convoys is based on considering several criteria such as the position, the movement direction of vehicles and the relative speed inside the convoy. The convoy is formed while taking into account the quality of the connections between the nodes and its maximum length in order to form convoys that are as stable as possible. We have proposed two merging and convoy division algorithms. The simulations we have carried out via a vehicular network simulator in a highway context have allowed us to study the distributions of the size and the length of the convoys. They have shown good stability for the convoys of 2 km in length. Since the vehicles are regrouped according to their speed, they remain together and there are few convoy breaks. Our next stage will be to complete this study with an implementation of CONVOY on a real platform, and then to carry out tests on the road, in order to test our simulation results in the field.

3.8. Bibliography

[AGA 09] AGARWAL R., MOTWANI M., "Survey of clustering algorithms for MANET", *Computing Research Repository*, vol. abs/0912.2303, 2009.

[ALM 10] ALMALAG M., WEIGLE M., "Using traffic flow for cluster formation in vehicular ad hoc networks", *Proceedings of the IEEE 35th Conference on Local Computer Networks (LCN 2010)*, Denver, CO, pp. 631–636, October 2010.

[BAR 04] BARR R., HAAS Z. J., Jist/Swans: Java in simulation time and scalable wireless ad hoc network simulator, 2004.

[BAS 99] BASAGNI S., "Distributed clustering for ad hoc networks", *Proceedings of the 4th International Symposium on Parallel Architectures, Algorithms, and Networks, (I-SPAN '99)*, Perth, Australia, 1999.

[BAS 01] BASU P., KHAN N., LITTLE T., "A mobility based metric for clustering in mobile ad hoc networks", *Proceedings of the 21st International Conference on Distributed Computing Systems Workshops (ICDCS 2001 Workshops)*, Phoenix, AZ, pp. 413–418, 2001.

[BLU 03] BLUM J., ESKANDARIAN A., HOFFMAN L., "Mobility management in IVC networks", *Proceedings of the IEEE Intelligent Vehicles Symposium*, Baltimore, MD, pp. 150–155, 2003.

[BON 07] BONONI L., DI FELICE M., "A Cross Layered MAC and clustering scheme for efficient broadcast in VANETs", *Proceedings of the IEEE Internatonal Conference on Mobile ad hoc and Sensor Systems (MASS 2007)*, Pisa, Italy, pp. 1–8, 2007.

[CAP 06] CAPPELLE C., POMORSKI D., YANQIN Y., "GPS/INS data fusion for land vehicle localization", *IMACS Multiconference, Computational Engineering in Systems Applications (CESA'2006)*, Beijing, China, pp. 21–27, 2006.

[DAS 05] DAS S.M., PUCHA H., HU Y.C., "Performance comparison of scalable location services for geographic ad hoc routing", *Proceedings of the 24th Annual Joint Conference of the IEEE Computer and Communications Societies (INFOCOM 2005)*, vol. 2, Miami, FL, 2005.

[DEA 09] DEAN A., BRENNAN S., "Terrain-based road vehicle localization on multi-lane highways", *Proceedings of the American Control Conference (ACC'09)*, Saint Louis, USA, pp. 707–712, 2009.

[DUC 07] DUCOURTHIAL B., KHALED Y., SHAWKY M., "Conditional transmissions: performance study of a new communication strategy in VANET", *IEEE Transactions on Vehicular Technology*, vol. 56, no. 6, pp. 3348–3357, 2007.

[FAN 05] FAN P., HARAN J.G., DILLENBURG J., *et al.*, "Cluster-based framework in vehicular ad hoc networks", *Proceedings of the 4th International Conference on Ad-Hoc, Mobile, and Wireless Networks (ADHOC-NOW'05), LNCS 3738*, Springer, New York, pp. 32–42, 2005.

[FAN 07] FAN P., MOHAMMADIAN A.K., NELSON P.C., *et al.*, "A novel direction based clustering algorithm in vehicular ad hoc networks", *Proceedings of the TRB 86th Annual Meeting Compendium of Papers CD-ROM*, Transportation Research Board Business Office, Washington DC, 2007.

[GHO 08] GHOSH R., BASAGNI S., "Mitigating the impact of node mobility on ad hoc clustering", *Wireless Communications and Mobile Computing*, vol. 8, no. 3, pp. 295–308, 2008.

[GUN 07] GUNTER Y., WIEGEL B., GROSSMANN H., "Cluster-based medium access scheme for VANETs", *Proceedings of the 10th International IEEE Conference on Intelligent Transportation Systems (ITSC'07)*, Seattle, WA, pp. 343–348, 2007.

[JAY 12] JAYASUDHA K., CHANDRASEKAR C., "Hierarchical clustering based greedy routing in vehicular ad hoc networks", *European Journal of Scientific Research (EURASIP)*, vol. 67, no. 4, pp. 580–594, 2012.

[LEE 10] LEE K.C., LEE U., GERLA M., "Survey of routing protocols in vehicular ad hoc networks.", in WATFA M., (ed.), *Advances in Vehicular Ad-Hoc Networks: Developments and Challenges*, Chapter 8, IGI Global, pp. 149–170, 2010.

[LIN 06] LIN L., OSAFUNE T., LENARDI M., "Floating car data system enforcement through vehicle to vehicle communications", *Proceedings of the 6th International Conference on ITS Telecommunications*, Chegdu, China, pp. 122–126, 2006.

[MAB 06] MABIALA M., BUSSON A., VÈQUE V., "Analyse du trafic et du routage dans un réseau ad hoc de véhicules", *Colloque Francophone sur l'Ingénierie des Protocoles (CFIP)*, Tunisia, 2006.

[MAB 07] MABIALA M., BUSSON A., VÉQUE V., "Performance evaluation of VANET under realistic vehicular traffic assumption", in APPERT-ROLLAND, CÉCILE *et al.*, eds., *Traffic and Granular Flow (TGF'07)*, Springer Verlag, pp. 739–744, 2007.

[MAS 11a] MASLEKAR N., BOUSSEDJRA M., MOUZNA J., *et al.*, "VANET based adaptive traffic signal control", *Proceedings of the IEEE 73rd Vehicular Technology Conference (VTC-2011 Spring)*, Budapest, Hungary, 2011.

[MAS 11b] MASLEKAR N., MOUZNA J., LABIOD H., *et al.*, "Modified C-DRIVE: clustering based on direction in vehicular environment", *Proceedings of the IEEE Intelligent Vehicles Symposium (IV)*, Baden-Baden, Germany, pp. 845–850, 2011.

[MEN 07] MENOUAR H., LENARDI M., FILALI F., "Movement prediction-based routing (MOPR) concept for position-based routing in vehicular networks", *Proceedings of the IEEE 66th Vehicular Technology Conference (VTC-2007 Fall)*, Baltimore, MD, pp. 2101–2105, 2007.

[NIU 06] NIU X., TAO Z., WU G., *et al.*, "Hybrid cluster routing: an efficient routing protocol for mobile ad hoc networks", *Proceedings of the IEEE International Conference on Communications (ICC)*, Istanbul, Turkey, 2006.

[NOC 03] NOCETTI F. G., GONZALEZ J. S., STOJMENOVIC I., "Connectivity based k-Hop clustering in wireless networks", *Telecommunication Systems*, vol. 22, nos. 1–4, pp. 205–220, 2003.

[OHT 03] OHTA T., INOUE S., KAKUDA Y., "An adaptive multihop clustering scheme for highly mobile ad hoc networks", *Proceedings of the 6th International Symposium on Autonomous Decentralized Systems (ISADS'03)*, Pisa, Italy, pp. 293–300, 2003.

[OHT 07] OHTA T., MURAKAMI N., KAKUDA Y., "Performance evaluation of autonomous clustering for hierarchical routing protocols in mobile ad hoc networks", *Proceedings of the 27th International Conference on Distributed Computing Systems Workshops (ICDCS 2007 Workshops)*, Toronto, Canada, p. 56, 2007.

[OHT 12] OHTA Y., OHTA T., KAKUDA Y., "An autonomous clustering-based data transfer scheme using positions and moving direction of vehicles for VANETs", *Proceedings of the IEEE Wireless Communications and Networking Conference (WCNC 2012)*, pp. 2927–2931, 2012.

[PET 11] PETRESCU A., BOC M., IBARS C., "Joint IP networking and radio architecture for vehicular networks", *Proceedings of the 11th International Conference on ITS Telecommunications*, Saint Petersburg, Russia, pp. 230–236, 2011.

[RAW 08] RAWASHDEH Z.Y., MAHMUD S.M., "Media access technique for cluster-based vehicular ad hoc networks", *Proceedings of the IEEE 68th Vehicular Technology Conference (VTC-2008 Fall)*, Calgary, Canada, 2008.

[RAW 09] RAWSHDEH Z.Y., MAHMUD S.M., "Toward strongley connected clustering structure in vehicular ad hoc networks", *Proceedings of the IEEE 70th Vehicular Technology Conference Fall (VTC-2009 Fall)*, Anchorage, AK, 2009.

[REM 11] REMY G., SENOUCI S.-M., JAN F. *et al.*, "LTE4V2X – impact of high mobility in highway scenarios", *Proceedings of the Global Information Infrastructure Symposium (GIIS)*, Da Nang, Vietnam, pp. 1–7, 2011.

[REM 12] REMY G., SENOUCI S.-M., JAN F., *et al.*, "LTE4V2X – Collection, dissemination and multi-hop forwarding", *Proceedings of the IEEE International Conference on Communications (ICC)*, Ottawa, Canada, 2012.

[SAN 06] SANTOS R., EDWARDS A., "A reactive location routing algorithm with cluster-based flooding for inter-vehicle communication", *Computación y Sistemas*, vol. 9, no. 4, pp. 297–313, 2006.

[SCH 06] SCHMILZ R., LEIGGENER A., FESTAG A., *et al.*, "Analysis of path characteristics and transport protocol design in vehicular ad hoc networks", *Proceedings of the IEEE 63rd Vehicular Technology Conference, (VTC-2006-Spring)*, vol. 2, Melbourne, Australia, pp. 528–532, 2006.

[SHE 09] SHEA C., HASSANABADI B., VALAEE S., "Mobility-based clustering in VANETs using affinity propagation", *Proceedings of the IEEE Global Telecommunications Conference (GLOBECOM 2009)*, Hawaii, USA, pp. 1–6, 2009.

[SOU 10] SOUZA E., NIKOLAIDIS I., GBURZYNSKI P., "A new aggregate local mobility (ALM) clustering algorithm for VANETs", *Proceedings of the IEEE International Conference on Communications (ICC)*, Cape Town, South Africa, pp. 1–5, 2010.

[TAL 06] TALEB T., OCHI M., JAMALIPOUR A., *et al.*, "An efficient vehicle-heading based routing protocol for VANET networks", *Proceedings of the IEEE Wireless Communications and Networking Conference (WCNC'06)*, Las Vegas, NV, pp. 2199–2204, 2006.

[WIS 07] WISITPONGPHAN N., BAI F., MUDALIGE P., *et al.*, "Routing in sparse vehicular ad hoc wireless networks", *IEEE Journal on Selected Areas in Communications*, vol. 25, no. 8, pp. 1538–1556, 2007.

[YU 05] YU J.Y., CHONG P.H.J., "A survey of clustering schemes for mobile ad hoc networks", *IEEE Communications Surveys and Tutorials*, vol. 7, no. 1, pp. 32–48, 2005.

Chapter 4

Complementarity between Vehicular Networks and LTE Networks

4.1. Introduction

Our use of communication means has become more and more mobile, and this is why, in the last few years, we have witnessed the creation of new communication and information technologies (CIT) that provide Internet connectivity everywhere and all the time (at home, at work, during our commute as well as our leisure travels). We spend more and more of our time in a form of transport, whether it is our personal car or public transport. The concept of vehicular *ad hoc* networks (VANETs) was born as a result of these developments. VANETs are being paid increasingly more attention from the academic field as well as the industrial field. Today, for example, several organizations support research, development and standardization activities aiming to provide a large variety of services, from road safety

---

Chapter written by Guillaume RÉMY, Sidi-Mohammed SENOUCI, François JAN and Yvon GOURHANT.

to traffic optimization, including driver and/or passenger entertainment. However, numerous challenges in this type of network still remain undiscovered. When compared to other traditional wireless networks, this type of network does indeed have particular characteristics, such as the relatively high speed of the vehicles and the extremely dynamic topology of the network. These characteristics must be considered throughout the development of new protocols for VANETs. To do this, several works propose that we automatically organize the network [CHE 09b, CHE 09a, CHI 02, YIZ 10]. Most of these architectures are based on the dynamic, decentralized creation of virtual vehicle clusters. However, with a highly dynamic network such as VANET, a decentralized approach becomes inappropriate, because it quickly generates a strong overload in the network.

In this chapter, we propose LTE4V2X (long-term evolution for V2X communications), an innovative solution for a "centralized" organization of vehicle networks by using the 4G LTE mobile network. We have chosen LTE for various reasons:

– LTE can largely support the traffic induced by our protocol (this hypothesis is verified by simulations).

– The coverage of an eNodeB is approximately 1 km^2 in a city and is thus well adapted to the organization and management of a vehicle network: approximately 400 vehicles need to be managed in a large city.

– The use of a centralized mechanism seems to be intuitively better than the use of a decentralized mechanism, since an eNodeB (the equivalent of a base station (BS) in an LTE network) has a global view of its coverage area and can thus improve the management and maintenance of the clusters.

– The majority of proposed VANET architectures, such as continuous air interface for long and medium distance (CALM, ISO TC 204 (ITS) Working Group 16), presuppose the existence of multiple interfaces in the vehicle, including 802.11p and LTE.

The final objective of LTE4V2X is to create an organized structure in order to facilitate the development of applications and protocols for vehicular networks. Among these applications, we have looked into information collection and dissemination applications in a VANET. Indeed, the vehicles become more and more sophisticated and gather information not only from the operational state, but also from their general environment, via internal or external sensors, radars or Global Positioning Systems (geographical position, speed, acceleration, temperature, road conditions, battery level for an electrical car, tire inflation level, etc.). Such information offers service providers an opportunity to offer a range of various applications to the drivers and/or passengers of a vehicle. Gathering such a volume of information raises several problems that LTE4V2X is indeed capable of resolving efficiently.

LTE4V2X uses the two technologies 802.11p and LTE at the same time in order to provide an efficient way of periodically collecting data from vehicles, and sending them to a main server. We manage to make these two types of network interact with one another in order to become complementary and to mutually optimize each other. Thus, LTE4V2X uses the eNodeB to organize the vehicular network in a centralized manner. Its functioning draws from the centralized dynamic clustering (CDC) protocol [BAJ 09], a centralized clustering protocol for wireless sensor networks. This protocol not only dynamically creates clusters, but also changes the "clusterhead" (CH) with each round so as to avoid the battery of the CH running out prematurely (the notions of "clusterhead" and "round" will be explained in the following sections). Since our architecture uses vehicles, and not wireless sensors, as nodes in the network, the energy consumption does not impose as many constraints. Consequently, CDC has been reinvented by taking into consideration the constraints specific to vehicular networks (fast movement, quasi-unlimited electric energy, etc.). LTE4V2X was thus developed starting from this initial idea, and it represents a complete architecture of a centralized organization of a vehicular network.

We have studied the performances of LTE4V2X against a decentralized management approach of the vehicular network for a well-known application: the floating car data (FCD) application, or floating vehicular data. The FCD applications are based on localization data collection, on speed and direction, in order to feed a traffic management server. The use of these data will, among other things, help drivers avoid traffic jams or magnetic loops integrated in the road, without requiring that any supplementary material be connected to the road. The FCD version used in this architecture is DFCD. DFCD is the decentralized FCD version, in the sense that the information that needs to be collected (position, speed, direction, etc.) is local to each vehicle, and each vehicle accesses its data before transmitting them via the network. We have also studied the performances of our solution for useful information dissemination applications (alerts, traffic info, etc.) at the destination of a vehicle network.

This chapter is structured into six parts. In section 4.2, we will present a summary of the actual knowledge and of the work that has already been done in the field of vehicular network organization and of the constraints specific to VANETs. After making a general presentation of the proposed architecture LTE4V2X in section 4.3, the protocol LTE4V2X-C (information collection protocol based on LTE4V2X) will be detailed in section 4.4. Section 4.5 will detail the LTE4V2X-D protocol (information dissemination protocol, based on the same architecture). In section 4.6, we study the performance of our two protocols. Finally, section 4.7 will conclude the chapter and introduce several guidelines for future research.

4.2. State of the art

The more particular characteristics of vehicular networks, which are related to the relatively high speed of the vehicles and the highly dynamic topology that results from it, make the design of new valid solutions quite difficult in these networks. Since one of the main objectives of a VANET is to provide road safety services, the inter-vehicle communications must indeed be conceived with the goal of being as valid and safe as possible. We have previously looked into several recent works [BON 07, CAR 10, DUR 10], studying the impact of this high mobility on the connectivity in these networks. In [DUR 10], the authors have carried out an analytical study for an interurban scenario (highways) and proposed a means of predicting 1-connectivity in the network, that is the probability of any vehicle being able to communicate with any other vehicle in multi-hops. They introduce a speed parameter, namely the equivalent speed. This parameter, which depends on the average and standard deviation of the speed of the vehicles, has the peculiarity of being able to create a relation between the mobility of vehicles and the connectivity within VANET: when the equivalent speed decreases, the connectivity in VANET increases. Thus, the authors were able to prove that the connectivity decreases when the average speed increases, and that the connectivity increases when the standard deviation of the speed of vehicles increases. In [CAR 10], the authors present a model for estimating the duration of the connectivity in a multi-hop VANET and the size of each group of vehicles that can communicate with each other on a highway. The authors give, among other things, the time during which the relay chain is established on the highway, as well as the packet loss rate depending on the distance between the source and the destination. They show, for example, that the packet loss rate remains lower than 1% when the distance is lower than 250 m (which corresponds to the range of 802.11p, when we only have direct communications), then increases linearly up to 500 m. After 500 m, the packet loss remains linear, but the multi-hop communications slow down the increase (the slope of the packet loss rate decreases). The authors have also developed a model that allows us to determine the density function of the probability of the connectivity period in the case of an interurban topology.

Next, we looked at the existing works on the organization of vehicle networks. Many of these works propose automatic organization architectures [BON 07, CHA 97, CHE 09a, KAI 11, WAN 08]. These architectures use the properties of the nodes to their advantage in order to provide a global virtual structure that allows for the self-organization of the network. In general, the architecture is sufficiently autonomous and dynamic in order to be able to manage local variations. Typically, in the case of vehicle networks, this often leads to a decentralized organization using clusters and/or a virtual backbone. The principle of this solution is to build a subnetwork of the best interconnected nodes. The only main constraint of this solution is the judicious choice of the backbone members (BMs) in order to avoid the loss of the interconnection between them. The second self-organizing structure is the clustering. This consists of partitioning the network into virtual areas called clusters. Each cluster has a CH, and potentially one or several members of the cluster. In general, the members of a cluster have common characteristics, such as speed, coordinates and/or similar directions. The solutions based on clusters represent a valid approach for the information-collecting applications. The cluster-based structure is often used as a support for the backbone structure. In [CHA 97], the authors draw inspiration from the organization of the mobile network in adjacent cells in order to divide the service area into sub-service areas (SSAs). A BS is installed in each SSA. The total coverage area of SSA is larger than the coverage area of a BS. Thus, a self-organization mechanism is executed in each SSA in order to ensure the continuity of the communication between the BS and the mobile station (MS), which is outside of its coverage area. In this method, several MS can be selected as relay MS, and thus support two radio communication channels: one for controlling the establishment of a link, and one for the data exchange. In [BON 07], the authors propose a dynamic backbone assisted Medium Access Control (MAC) (DBA-MAC) in the context of VANETs, which is a proactive distributed method for forming a backbone in a dynamic fashion, with the aim of disseminating an alert message to a group in a risk area. To create the backbone, a node begins by electing itself as a member of the backbone, then disseminates a message in order to give it the necessary impulse for the creation of the backbone. Then, all the receptors enter the distributed phase of access to the medium based on the contention mechanism, in order to choose the next BM. Since a reactive mechanism of repairing the backbone would require us to introduce a detection method of the state of the links and of the overload in the network, DBA-MAC updates the backbone in a proactive manner. In [WAN 08], the authors propose a cluster-based self-organizing protocol (CSP), a proactive self-organization architecture of a vehicular network, based on the geographical clustering in order to ensure a permanent self-organization of the entire network. CSP adapts itself to the characteristics of the vehicular network and allows us to improve the connectivity between the vehicles or between the vehicles and the infrastructure.

However, we can easily suppose nowadays that the vehicles will have multiple network interfaces, which will allow for a connectivity to take place at the level of the 802.11p network as well as in the mobile network (UMTS/LTE). In a recent work [BEN 11, TAL 10], the authors propose a decentralized clustering algorithm for a heterogeneous mechanism composed of a vehicular network and a mobile network. In the proposed architecture, the mobile network is not only used for uploading the data coming from the vehicle, but also plays an indirect role in forming the clusters: the relative directions of the vehicles in relation to the BS are considered in their cluster formation algorithm. Just as in the aforementioned works, the process of the creation and maintenance of clusters is decentralized. However, it might be interesting to take advantage of the presence of the mobile network in order to help form clusters and to optimize cluster formation. The BS (NodeB for a Universal Mobile Telecommunications System (UMTS) network or eNodeB for an LTE network) can have a global vision of the vehicular network and can be used to organize it in a centralized manner. The centralized approach allows the cluster manager (NodeB or eNodeB) to have a global view of the road traffic, and therefore to potentially create better clusters than a centralized approach that only has a limited and local view of the network. In addition, this global view allows us to easily detect the changes in the topology of the vehicular network caused by the movements of the vehicles. Thus, the LTE4V2X architecture that we propose uses this centralized approach for creating and maintaining clusters.

In the following sections, we will detail our solution of centralized organization of a VANET that benefits from the existence of an LTE network.

4.3. General description of the proposed architecture

In this section, we present LTE4V2X, a new architecture for organizing a vehicular network of type 802.11p while using an operating network of type 4G LTE. Indeed, we suppose that all the vehicles have two communication interfaces: an LTE interface and a 802.11p interface. This presupposition is completely acceptable in the given context. The organization of the vehicular network consists of organizing vehicles in a set of clusters generated by the eNodeB (the clusters can equally be generated by a distant server when it is no longer possible to modify the eNodeB). We demonstrate the feasibility of our architecture for two types of application: an information collection application and an information dissemination application. First of all, we begin by presenting the architecture in the case of a complete LTE coverage, then in the case of a partial coverage (in the presence of a tunnel, for example). Then, we present the two protocols by using this architecture: LTE4V2X-C for an application of information collection, and LTE4V2X-D for an information dissemination application.

4.3.1. *Network organization mechanisms for areas completely covered by LTE*

In the regions completely covered by the mobile network, LTE4V2X organizes the vehicular network in a set of clusters, in a centralized manner. The size of a cluster is not larger than the range of the 802.11p, which ensures that all of the nodes of a cluster could join whichever other node of the same cluster, without needing multi-hop communications. Each eNodeB manages the set of clusters that are within its range (the LTE coverage has a diameter of approximately four times greater then the one in the coverage area 802.11p). In each cluster, a CH is elected (see Figure 4.1). The CH has the responsibility of collecting and sending its data as well as the data sent out by the members of the cluster to the eNodeB via the LTE network. The simple members of the cluster send their data uniquely to their CH via the network 802.11p.

4.3.2. *Network organization mechanisms for areas that are not completely covered by LTE*

In the cells that are partially covered by the mobile network, for example in the presence of a tunnel, we introduce the conditional multi-hop mechanism: when the vehicle goes into the tunnel, it passes into multi-hop mode, and uses a protocol that is independent of the standard LTE4V2X protocol. The protocol used in these areas that are not covered by the mobile network is a protocol of decentralized self-organization, based on clustered gathering protocol (CGP) [SAL 09]. CGP uses a topology of fixed clusters for organizing the network: the road is divided into segments of equal length, and each segment corresponds to a cluster. The segments have a length of 150 m, with the aim of ensuring that the CH can join the CG of the adjacent segments by using a direct (without relay) wireless communication. Such a mode will improve the performances of the applications based on the gathering and the dissemination of data, in particular in the areas that are partially not covered by the LTE mobile network. For the applications based on the gathering of information, such as FCD, CGP uses a periodic scheme of rounds divided into three phases. These phases are the election of the CH (CH election), data aggregation and data collection. For the applications based on information dissemination, we use the CH election phase, but then the packets that must be disseminated are freely disseminated according to the scheme that will be presented further on. As far as this presentation is concerned, and since it corresponds to the scenario used in the simulations, we will discuss the tunnel here in order to describe the areas that do not receive the LTE signal.

4.3.3. *Information collection application: LTE4V2X-C*

The proposed LTEV2X architecture can easily be used by an information collection application. This has given birth to a new collection protocol that we have called LTE4V2X-C. In the latter protocol, all of the vehicles can be elected as a CH. The CHs aggregate the data from the cluster members before sending them to the eNodeB (see Figure 4.1). This allows us to reduce the use of the LTE network by avoiding sending useless data (e.g. when the direction and speed of a vehicle has remained unchanged, the CH can avoid sending the data of this node to the eNodeB). The CH can also use a compression algorithm on the aggregated data in order to reduce the volume even more. The details of this protocol will be presented later on.

4.3.4. *Information dissemination application: LTE4V2X-D*

Just like the collection applications, the LTEV2X architecture can also be used by an information dissemination application. This has given birth to a new dissemination protocol that we have called LTE4V2X-D. Disseminating a piece of information in a given geographical area is one of the most important applications in a vehicular network (e.g. danger alert and traffic information), and is sometimes even critical. In its normal configuration, when all the vehicles are within the range of the eNodeB and can receive data from it, LTE4V2X-D uses direct dissemination via LTE. If not, it uses the previously presented multi-hop protocol.

4.4. Detailed description of the LTE4V2X-C protocol

In this section, we present the different mechanisms of our LTE4V2X-C protocol. We have chosen the FCD application as a collection application by using our organization architecture. After the initializing phase (see section 4.4.1), the system functions in a cyclical fashion (see Figure 4.2): each round contains three phases and is repeated indefinitely. These three phases are as follows:

– the configuration phase, when the eNodeB creates and updates the clusters;

– the advertisement phase (ADV), during which the CHs send a notification message in the VANET;

– the collection and aggregation phase, during which the cluster members send their FCD data to the CHs, and the CHs send the aggregated FCDs to the eNodeB.

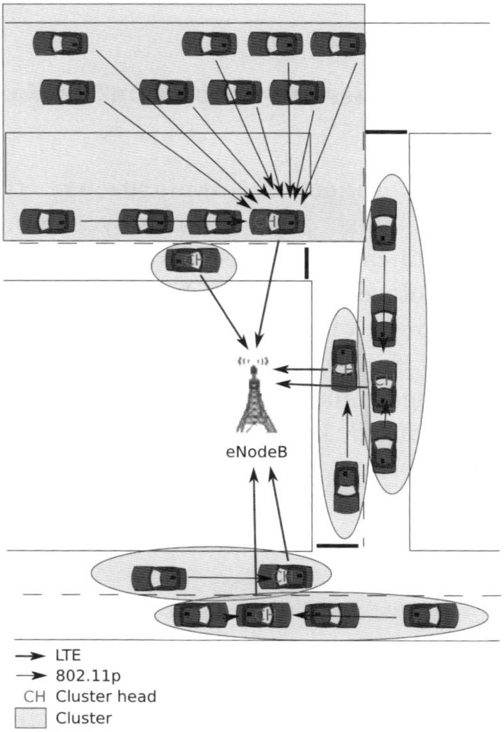

Figure 4.1. *LTE4V2X architecture for data collection*

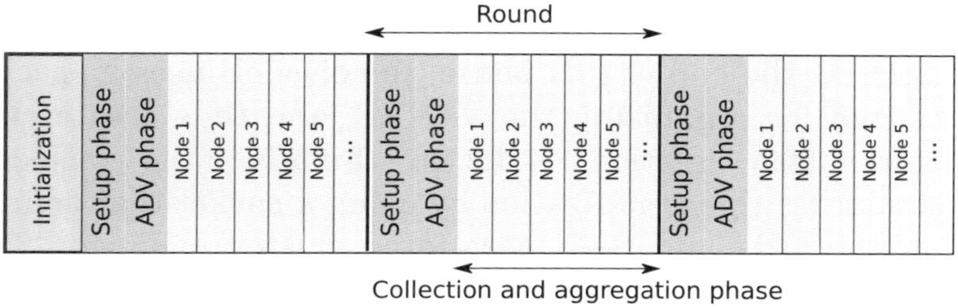

Figure 4.2. *Round phases*

Let us note that the nodes use the time division multiple access (TDMA) in each cluster in order to send their FCD data (clock, position, speed and direction) to the CH.

First, we will detail the initialization phase that only appears at the beginning of the system. Then, we will describe how the organization into clusters is maintained by the eNodeB, and how the eNodeB manages the arrivals and departures of the vehicles from the network. Finally, we will explain the aggregation collection phase, where the eNodeB collects the FCD data of the vehicular network.

4.4.1. *Initialization phase*

During the initialization phase, the eNodeB receives the FCD data of all the vehicles directly via the LTE interface and thus without any intervention from the 802.11p network. This phase is executed only once, when the network starts, and allows the eNodeB to discover the positions of the vehicles before starting to create the clusters. In the beginning, the vehicular network is not organized and there are still no clusters. All of the vehicles send an identifier request by using the packet format described in Figure 4.3, until they obtain an identifier provided by the eNodeB. Then, the nodes send their FCD data (clock, position, speed and direction) to the eNodeB (see Figure 4.4), by using the same mechanism as if they entered the network (see section 4.4.2.1). At the end of this phase, the eNodeB uses a specific algorithm for organizing the VANET, and thus creates the clusters (see section 4.4.2.3). Finally, the eNodeB disseminates this organization by using a packet that defines all of the CHs and their associated cluster members. The format of this packet is described in Figure 4.5.

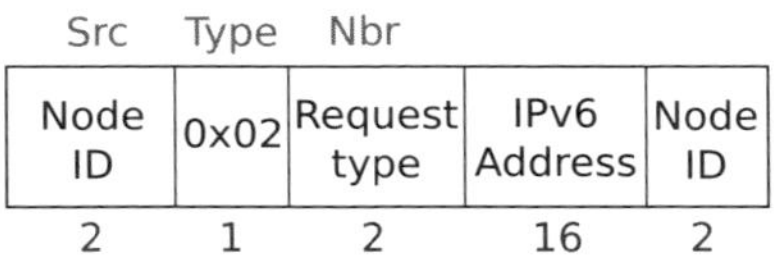

Figure 4.3. *Identification packet*

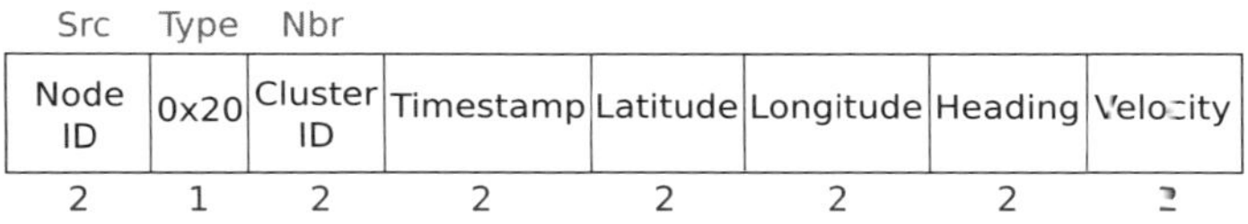

Figure 4.4. *FCD packet*

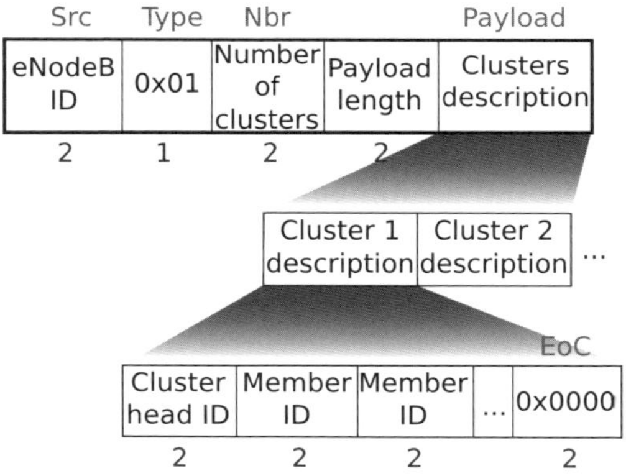

Figure 4.5. *Topology packet*

The identifier request messages are used for adding a new association to the address IPv6/ID, in order to suppress an existing association, or for demanding the address IPv6 that corresponds to a given identifier.

4.4.2. *Maintenance*

In this section, we detail how the LTE4V2X-C maintains the clusters with the help of the eNodeB: how the eNodeB manages the arrivals and departures of vehicles, what periodic controls are being effectuated and what control messages are being sent by the eNodeB.

4.4.2.1. *Arrival of new nodes*

The arrival of a new node in a cluster is pretty simple: first, the node that arrives waits for the reception of an advertisement message (ADV) of the VANET, of an update message of the eNodeB clusters or the timeout of a predefined delay. The reception of this message lets the new node know when the next collection stage and aggregation stages will commence. During this phase, the node will send its FCD data straight to the eNodeB, just like in the initialization phase (with a cluster identifier equal to 0). After that, the eNodeB will integrate the new node into the next configuration stage and will consequently send an update packet of the clusters.

4.4.2.2. *Departure of a node*

The behavior will obviously differ if the node leaving the network is a CH or a simple member of the cluster. When the departing node is a member of the cluster, its departure is detected one round after the exit of a node, at the latest, because the eNodeB can easily see that it has not received the FCDs of one of the members of a cluster. The departing node will be deleted from the list of nodes generated by the eNodeB after a certain number of consecutive rounds without receiving any FCDs from it. When the departing node is a CH, the process is different. If the departing CH is disconnected before sending the aggregated FCDs of the cluster, all the FCDs of the cluster in question will be lost for this round. The eNodeB will assume that all the nodes of the cluster are faulty (in case other nodes besides the CH had left the network), and will therefore put in several rounds in order to re-establish a normal situation for the members of the respective cluster.

4.4.2.3. *Periodical operations*

Periodical verifications are carried out by the eNodeB with each round, throughout the configuration phase. The eNodeB tries to create clusters that have the longest possible lifetime. The size of a cluster does not surpass the range of the 802.11p communications, in order to ensure that each node can directly join whichever other node of the cluster (in particular, the CH). The eNodeB creates clusters that contain the greatest possible number of nodes that go in the same direction (see Figure 4.1), as described by algorithm 4.1. This algorithm is periodically executed by the eNodeB with each round, throughout the configuration phase.

At the end of the configuration phase, the eNodeB disseminates an update packet of the clusters (see Figure 4.6). This update packet contains the commands for each cluster that needs to be modified. These commands can indicate the addition of a new node to the cluster, its retrieval, the identifier of the new CH of the cluster or even the change in the identifier of the cluster.

During the advertisement phase (ADV), the configuration of the clusters is disseminated in the VANET: each CH disseminates a short message that contains the future CH from its cluster and the number of nodes in the cluster (see Figure 4.7). This allows the members of the clusters to synchronize with their CH and to confirm that it is present and that it will be able to follow the data that are useful to the eNodeB.

Algorithm 4.1. *Clustering algorithm*

... $\rightarrow$ All nodes are considered as not assigned to a cluster
while Remaining unassigned nodes > 0 **do**
... $\rightarrow$ For each node, calculate the number of neighboring vehicles (at less than 300 m apart) that go in the same direction as the node
... $\rightarrow$ Create a new empty cluster
... $\rightarrow$ Assign the vehicle that has the most neighbors as well as all the neighbors at the new cluster
... $\rightarrow$ Determine the vehicle that has the best LTE reception in this new cluster, and select it as CH
end while

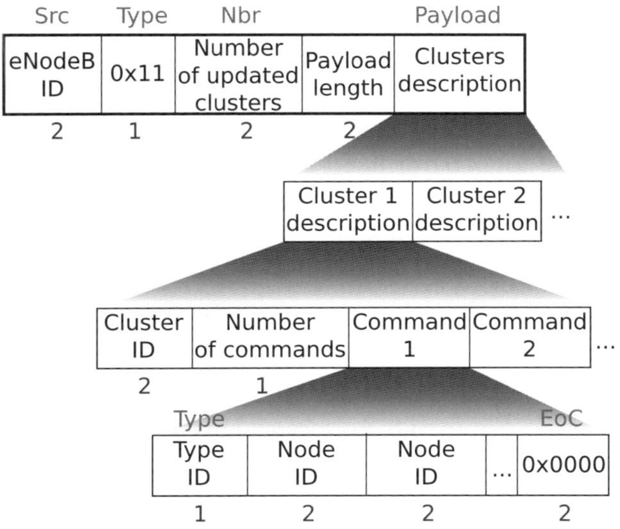

Figure 4.6. *"Cluster update" packet*

Src	Type	Nbr		
Node ID	0x40	Cluster ID	Next CH ID	Cluster size
2	1	2	2	2

Figure 4.7. *Advertisement packet in the VANET*

4.4.2.4. *Collection and aggregation phase*

Throughout this phase, each node sends its FCD data (clock, position, speed and direction) to its CH in its attributed time slot (see Figure 4.4). The time slots are calculated depending on the order used by the eNodeB in its last update message and on the identifier of the vehicles. If the node does not know the IPv6 address associated with the identifier of its CH, it sends an identifier request to the eNodeB (see section 4.4.1).

At the end of this phase, each CH sends the aggregated (and potentially compressed) data to the eNodeB (see Figure 4.8). The size of this packet depends on the chosen compression algorithm, on the aggregation process and obviously on the number of nodes in the cluster. For example, with 30 nodes in the cluster and a compression algorithm similar to the algorithm used by the tar-gzip format, the size of the packet reaches approximately 128 octets (without taking into account the aggregation process).

Src	Type	Nbr			
Node ID	0x22	Cluster ID	Size of compressed data	Size of uncompressed data	Compressed data
2	1	2	2	2	

Figure 4.8. *Aggregated FCD packet*

4.4.3. *Extension for the areas not covered by the LTE*

As presented in section 4.3, we have extended the LTE4V2X architecture to the case of areas that are not directly covered by the LTE network. In this case, we use a round scheme made of three phases, which repeat themselves indefinitely. These phases are the election of CH, the aggregation of data and the collection of data.

The mechanism for the election of the CH chooses the node that is the most advanced in the cluster. This is done not only in order to reduce the delay between the reception of data by the CH and their retransmission to the adjacent cluster, but also in order to forward the message more rapidly in case of empty clusters (store-and-forward). To achieve this end, CH_ANNOUNCE messages and a back-off time have been introduced. The CH_ANNOUNCE messages are short packets potentially diffused by all of the nodes throughout the CH election phase. The vehicles send the CH_ANNOUNCE packet starting from the beginning of the CH election phase, at the end of a delay calculated for each round by each vehicle, and what we call the back-off time. This is calculated in such a way that the vehicles that drive at a high speed and that are about to leave the cluster obtain very low back-off time (see [4.1]). When a node receives the message CH_ANNOUNCE from another node from its cluster, it cancels the sending out of its own CH_ANNOUNCE (because this signifies that the back-off time of the node that has sent this message is lower than its own back-off, and therefore that the other node is a better candidate in the place of the CH). Thus, the CH-elected vehicle is the first one to send its CH_ANNOUNCE message. This election phase is also used by the nodes for detecting the CHs of neighboring clusters, which will be very useful for the last phase of the round.

$$t_{\text{back-off}}(i) = \frac{\text{segment length} \cdot \exp\left(-\frac{\text{speed}}{\text{max speed}}\right)}{2 \cdot \text{segment length} - \text{distance}(\text{end segment}, \text{position}(i))} \quad [4.1]$$

During the aggregation phase, each vehicle that is not a CH sends its FCD data to its CH. It uses its back-off time (calculated in the CH election phase) in order to avoid the collisions of FCD packets between the members of the cluster. Then, each CH aggregates and compresses the received FCD data.

Throughout the collection phase, the CHs that are in the tunnel try to send the aggregated FCD data to the eNodeB by using multi-hop communications (see Figure 4.9). They use the neighboring CHs detected in the CH election phase in order to have an overview of the local topology of the network, and thus to try to choose the best CH for channeling their data. Each packet containing aggregated FCDs is channeled toward the interior of the tunnel via multi-hop communications between the CHs, until it reaches one of the extremities of the tunnel. The version of this protocol that we have used in our simulations supposes that the vehicles know beforehand the areas that are not covered by the mobile network (we can realistically suppose that the tunnels, for instance, are registered in a database and signaled to vehicles by the eNodeB). With this information, the nodes can choose more efficiently the CH that will forward their messages. Thus, in a tunnel, if a node knows that it has not yet reached the middle of the tunnel, it will try to send its message backward. However, we can also imagine a version of this protocol where this non-coverage zone is not known, and where the vehicles will therefore send their messages in multiple directions, with the aim of making sure that at least one of the messages reaches the edge of the area.

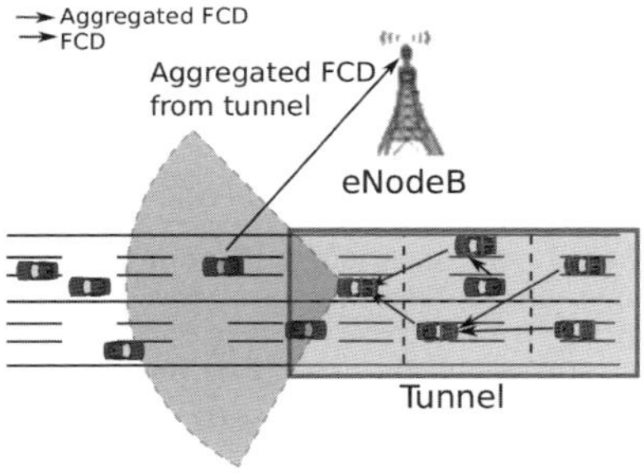

Figure 4.9. *Data collection in multi-hop*

When the packets reach an end of the tunnel, there is a final stage before the eNodeB can recover the collected FCD data. It must interconnect the centralized LTE4V2X-C with this multi-hop extension based on the CGP, so that the eNodeB receives these FCD and takes them into account on the same basis as those coming from the centralized LTE4V2X-C. To do this, we will consider that when a vehicle is near an extremity of the tunnel, it will disseminate its blindly collected FCD data. This will allow the nearest nodes that are covered by the LTE network to receive these FCDs that come from the tunnel, and to forward them immediately to the eNodeB.

4.5. A detailed description of the LTE4V2X-D protocol

In this section, we present the different mechanisms of our LTE4V2X-D dissemination protocol. Because of the critical aspect of the majority of the dissemination applications, they do not respect round phases like those used by LTE4V2X-C. The eNodeB diffuses directly the notification message to all its vehicles, accompanied by the head of the packet that contains the area concerned by the notification (see

Figure 4.10). Thus, all the vehicles that drive in the area concerned will receive the message and will consider it after reading the head of the packet. The eNodeB redisseminates the message periodically, with the aim of ensuring that the nodes arriving in the area are notified of this.

Src	Type	Nbr				
Node ID	0x40	Next forwarder	Notification flags	Notification position	Notification range	Notification timestamp
2	1	2	1	6	2	8

Figure 4.10. *Notification packet*

If the tunnel is concerned with the notification (the area that needs to be notified covers all, or just a part of the tunnel), all of the vehicles that are close to the tunnel disseminate the notification in the vehicle network. Then, the vehicles that are in the tunnel forward the information by using the multi-hop extension based on CGP and that we have presented previously. Contrary to the collection applications, we send the message that needs to be disseminated straight after the CH election phase. The message is then sent from CH to CH, while trying to notify as many nodes as possible at the passage (see Figure 4.11).

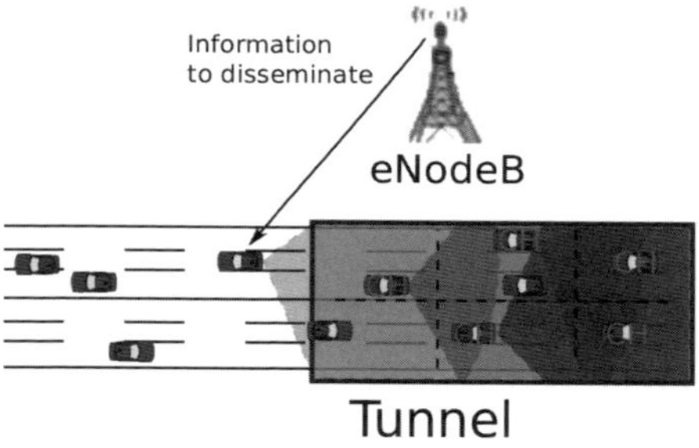

Figure 4.11. *Multi-hop dissemination*

4.6. Performance evaluation

This section presents the performance studies carried out on LTE4V2X, and specifically on the two associated implementations LTE4V2X-C and LTE4V2X-D. Section 4.6.1 introduces the simulation hypotheses and the different metrics used. Then, we present the results of the simulation and analyze them. Finally, we detail a short analytical study that we have resorted to for studying the effect of the handover (LTE cell change) on the collection protocol LTE4V2X-C.

4.6.1. *Hypotheses*

The simulations were carried out with the network simulator *NS-3* [FOU 12]. The proposed architecture LTE4V2X and the two associated protocols LTE4V2X-C and LTE4V2X-D were implemented above the layer IP. The parameters that all of these simulations have in common are listed in Table 4.1, and the parameters specific to the information collection application are listed in Table 4.2. We have carried out simulations for urban and interurban topologies (parameters that are also described in Table 4.1). Between each simulation case, we have varied either the number of vehicles, or the speed of the vehicles.

We have compared our protocols to a decentralized clustering protocol (DCP). In DCP, the VANET is self-organized in a decentralized manner. The vehicles calculate the clusters each second, and send an ADV packet in case they are elected CH. Then, the behavior varies depending on the type of application. For a collection application, all of the nodes that are not CH send their FCD data to their CH. The CH sends, after that, the aggregated data to the eNodeB. For a dissemination application, we use the clusters that have been created to directly send the information that needs to be disseminated in all the clusters. LTE4V2X-C and LTE4V2X-D have also been compared to their equivalent that does not benefit from multi-hop extensions (for the dissemination and the collection in the areas that are not covered by the LTE network). This has allowed us to determine the improvement of the performances of our protocols that this extension has brought about.

Parameter	Urban topology	Interurban topology
Surface simulated	1 km^2	8 km^2
Vehicle speed	20–50 km/h	100–400 km/h
Number of vehicles	100–400	100–300
Network physical layer	802.11p, LTE	
Maximum range of the 802.11p	300 m	
Simulated time	120 s	
Number of iterations per each simulation case	10	

Table 4.1. *Parameters connected to the topology*

Parameter	Value
Duration of a *round*	1 s
Duration of the initialization phase	10 s
Expiration of the control messages	after 1.5 s

Table 4.2. *Parameters for the FCD application*

For the simulations in an urban area, we have used an area of 1×1 km^2, which corresponds to the coverage area of an eNodeB in a city. We have chosen an urban topology, generated with *VanetMobiSim* [EUR 12] (see Figure 4.12). The eNodeB is placed at the center of this area. The vehicles drive with an average speed of 20 km/h in a residential area, 30 km/h in the center of the city and 50 km/h for the rest of the map.

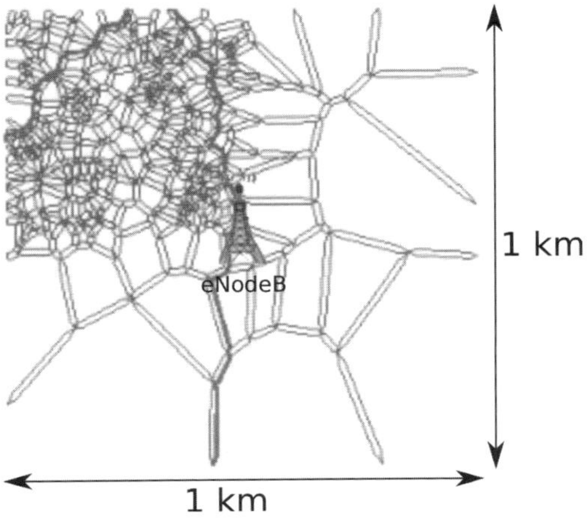

Figure 4.12. *Example of an urban topology*

The metrics used for evaluating the performances of the proposed protocols are the following:

– useful flow: this metric of an applicative level reflects the octets of useful data that were exchanged (FCD or information that needs to be disseminated);

– packet loss: this is the percentage of useful packets (FCD) sent by an element of the network but never received by the recipient;

– efficiency: this metric is introduced in order to reflect the capacity of the protocol to optimize the LTE bandwidth, while managing to keep reliable links within the vehicular network. It is given by the following equation:

$$E = 1 - \frac{1}{a} \qquad [4.2]$$

where:

$$a = \sqrt{\text{Number of vehicles per cluster} - 1} \cdot (1 - \text{Useful packet loss})^2 + 1$$

For the simulations in an interurban area, we have used an area of 8×1 km^2, 8×8 km^2 corresponding to the coverage area of a eNodeB in an interurban area. Two highways 2×3 ways cross this area, and the eNodeB is placed at the center (see Figure 4.13). We have added a tunnel of 4 km in length, represented by a rectangle in the figure, which corresponds to an area that is not covered by the LTE signal. This allows us to evaluate the performances of the multi-hop protocol. The area targeted by the dissemination protocol (the area to notify) is represented dark grey in the figure.

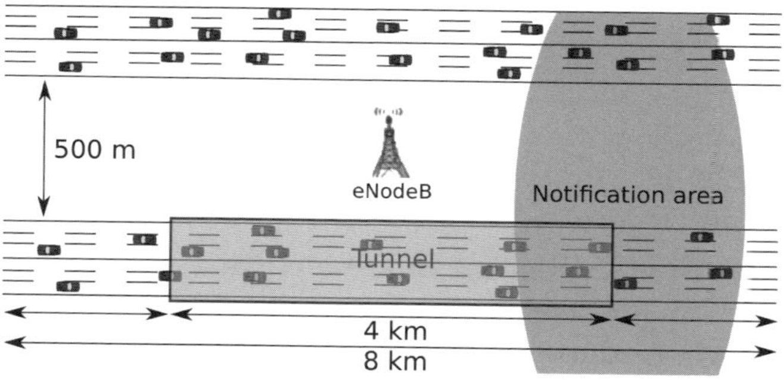

Figure 4.13. *Example of an interurban topology*

4.6.2. *The results of the simulation and their analysis*

In this section, we present the results of the simulations and analyze them, in order to be able to draw certain conclusions on the performances of LTE4V2X. Section 4.6.2.1 describes the results of the use of LTE4V2X for a LTE4V2X-C data collection application, and section 4.6.2.2 describes the results of the use of LTE4V2X for an LTE4V2X-D data dissemination application.

4.6.2.1. *LTE4V2X-C*

As explained previously, we have chosen FCD as a data collection application in order to assess our LTE4V2X-C protocol. We have first studied the evolution of the performance metrics depending on the density of the vehicles. Thus, we have obtained results for an urban topology (see Figures 4.14 and 4.15) and an interurban topology (see Figures 4.16 and 4.17). The multi-hop mechanisms at play in the areas not covered by the LTE network have not been tested for the simulations in an urban environment.

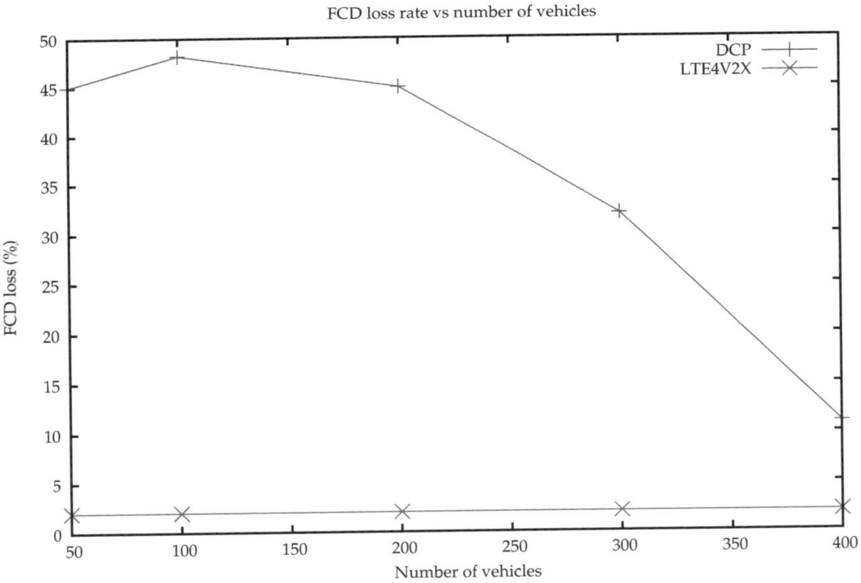

Figure 4.14. *Percentage of packet loss depending on the density of the vehicles for an urban topology*

Figures 4.14 and 4.15 focus on the simulations in an urban environment. Thus, Figure 4.14 represents the packet loss depending on the density of the vehicles in the case of an urban topology, for LTE4V2X-C and DCP. This figure clearly shows the extremely high packet loss that DCP suffers from. In fact, this decentralized approach creates less optimized clusters. The vehicles are not directly warned of the changes in topology. The simple cluster members can therefore send packets to non-existing nodes, and the CHs can miss the inclusion of a new node in the cluster. Moreover, with LTE4V2X, all of the nodes are synchronized with the help of the cluster update packet (sent periodically by the eNodeB in the LTE network). As far as DCP goes, it must try to synchronize the nodes without the help of a fixed infrastructure. Thus, this lack of synchronization, associated with the high dynamics of the network, provokes serious packet loss when the volume of the data exchanged in the VANET becomes significant. The packet loss observed in LTE4V2X remains constant, even at a high vehicle density. This can be explained by the fact that the loss is caused by the intrinsic dynamics of the network (the eNodeB cannot update the clusters more than once a second, and it can happen that the configurations need a higher updating frequency for the clusters).

Figure 4.15 represents the useful flow depending on the vehicle density in the case of an urban topology, for LTE4V2X-C and DCP. Two different curves were traced for each protocol, in order to be able to distinguish the flows of the LTW network from those of the vehicular network. This figure clearly shows that LTE4V2X generates a very low useful flow in the 802.11p network, and a relatively high useful flow in the LTE network. To understand this behavior, we must consider Figure 4.14. DCP indeed suffers from an extremely high packet loss, and we therefore obtain a low flow for DCP, given that a large part of the useful data is lost in the VANET. Then, the low value of the useful flow generated by LTE4V2X-C in VANET is explained by the fact that LTE4V2X-C creates clusters that have, on average, fewer vehicles than those in DCP. Therefore, fewer FCD data are exchanged in the vehicular network. LTE4V2X-C creates smaller clusters because its general vision of the VANET allows it to detect that certain large clusters are not optimized.

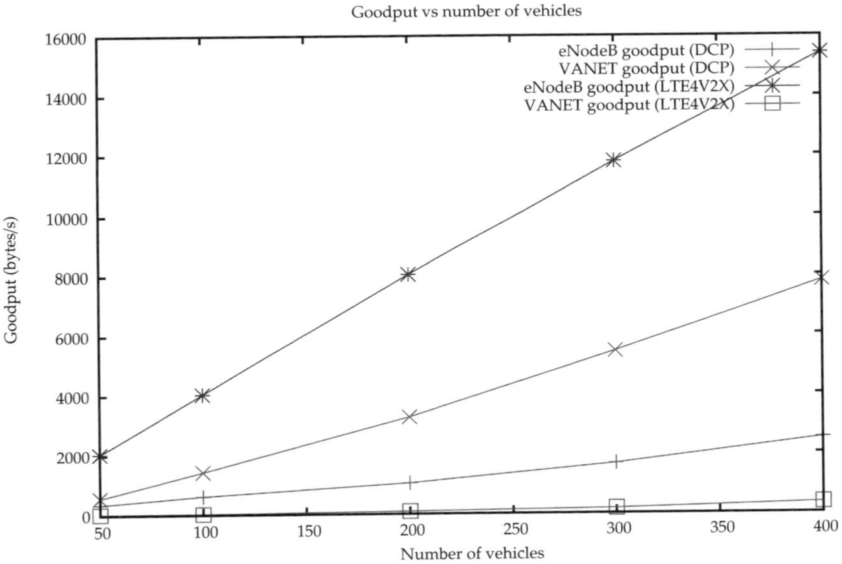

Figure 4.15. *Useful flow depending on the vehicle density for an urban topology*

Figures 4.16 and 4.17 focus on the simulations in an interurban environment. Figure 4.16 therefore shows the evolution of the packet loss depending on the density of the vehicles in the case of an interurban topology, for LTE4V2X-C, LTE4V2X-C without the multi-hop extension and DCP. It is worth noting on this curve that the packet loss for LTE4V2X-C slightly increases up to a density of 150 vehicles per km^2, then decreases when we continue to increase the number of vehicles. This is a direct effect of the multi-hop mechanism used by LTE4V2X-C. Starting from a certain vehicle density, we have a rather dense vehicle network for efficiently forwarding the information throughout the areas not covered by the LTE network. The efficiency of this mechanism increases with the number of vehicles, and therefore causes the packet loss percentage to plummet progressively. The packet loss for LTE4V2X-C without a multi-hop extension remains constant. Being already constant in an urban environment, it is logical that it remains equally constant in an interurban environment, with an offset of approximately 25%, which is easily explainable by the fact that the simulated highways are covered at 25% by a tunnel.

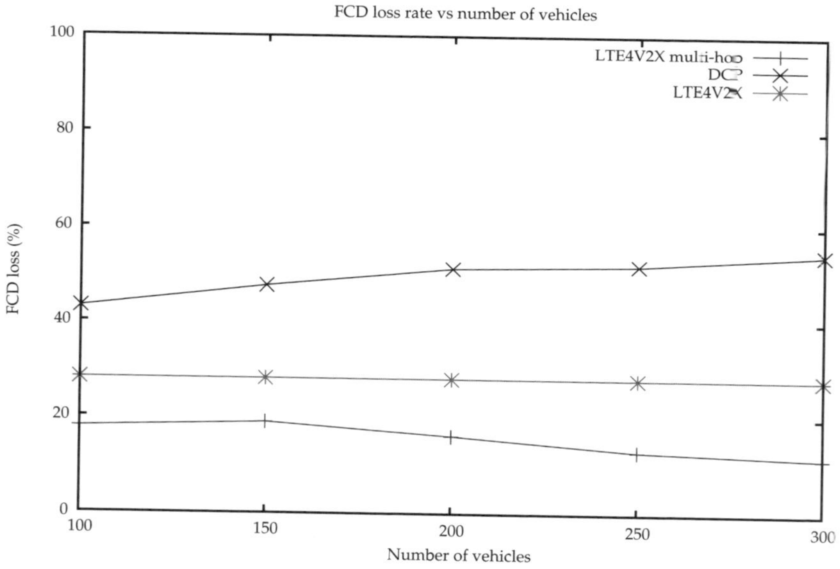

Figure 4.16. *Percentage of packet loss depending on the density of the vehicles for an interurban topology*

Figure 4.17 represents the efficiency depending on the density of the vehicles in the case of an interurban topology, for LTE4V2X-C, LTE4V2X-C without the multi-hop extension and DCP. As we might have expected after the previous results, LTE4V2X-C shows a higher efficiency than the other two protocols. The multi-hop mechanisms introduce a very slight overload in the network, but allow us to relay more useful data up to the eNodeB. LTE4V2X-C without a multi-hop extension is itself already better than DCP, because it combines an optimization of the use of the LTE channel and the clustering algorithms that allows a very slight loss of useful data. We can also note that the efficiency increases with the number of vehicles, because when we have a lot of vehicles per cluster, the collection is more optimized (we compress and aggregate more effectively).

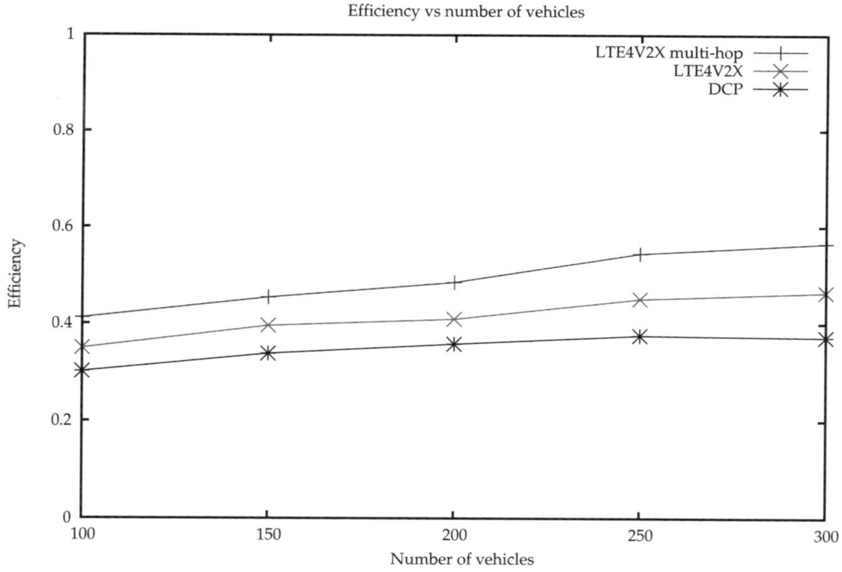

Figure 4.17. *Efficiency depending on the density of the vehicles for an inter-urban topology*

Having studied this evolution of the metrics depending on the density of the vehicles, we have moved on to simulations that allowed us to observe these same metrics depending on the speed of vehicles. To make sure we are truly able to study the effect of the speed on the protocols, we have chosen a model of mobility that is different for these simulations: we have chosen the ConstantSpeedMotion in VanetMobiSim, which consists of imposing a constant speed for each vehicle, for the entire duration of the simulation. Thus, we have obtained Figures 4.18 and 4.19.

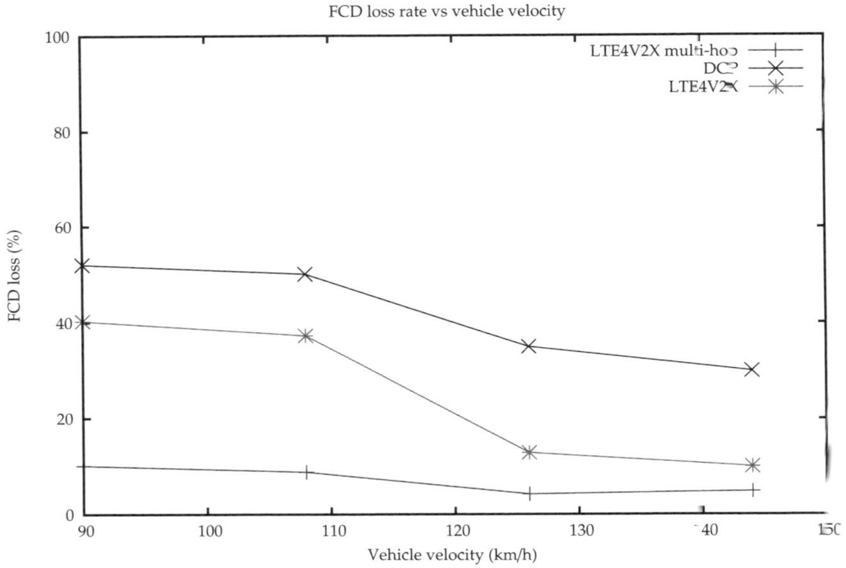

Figure 4.18. *Packet loss percentage depending on the speed of vehicles for an interurban topology*

Figure 4.18 shows the evolution of packet loss depending on the speed of vehicles in the case of an interurban topology, for LTE4V2X-C, LTE4V2X-C without multi-hop extension and DCP. We may clearly see that LTE4V2X-C benefits from the smallest packet loss, which could be explained by the fact that LTE4V2X-C creates, on the one hand, better clusters than DCP and, on the other hand, allows us to limit the packet loss in the areas not covered by the network, because of its multi-hop extension. The fast DCP decreases and LTE4V2X-C without multi-hop are caused by the tunnel: the higher the speed of the vehicles, the less time they spend in the tunnel, and this has a great bearing on the protocols that do not use multi-hop mechanisms. LTE4V2X-C also benefits from a decrease in packet loss when the speed of the vehicles increases, but in a less evident way, because its multi-hop extension allows it to be less affected by the tunnel.

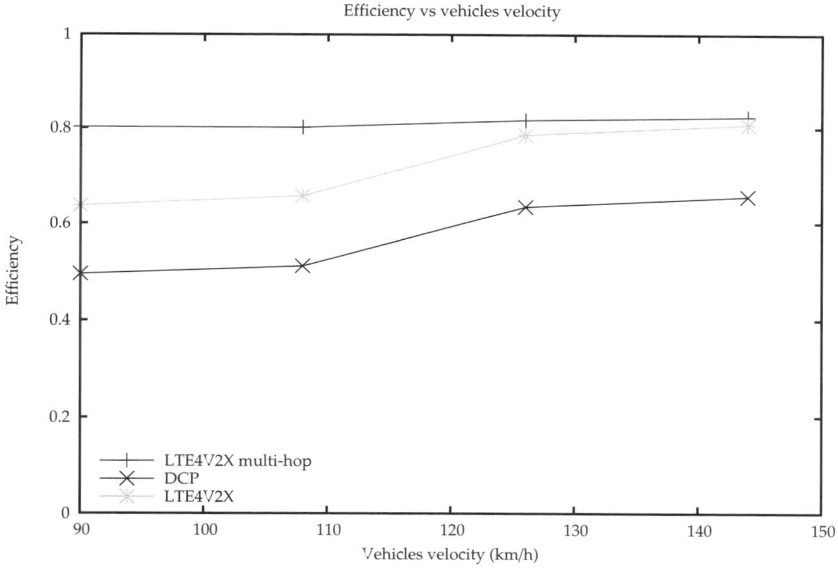

Figure 4.19. *Efficiency depending on the speed of vehicles for an interurban topology*

Figure 4.19 shows the evolution of the efficiency depending on the speed of the vehicles in the case of an interurban topology, for LTE4V2X-C, LTE4V2X-C without multi-hop extensions and DCP. We may note that, as expected, LTE4V2X-C is more efficient than LTE4V2X-C without multi-hops, which is itself better than DCP. We can explain this using the same reasons as those presented for Figure 4.17. In addition, we can note that the efficiency increases when the speed of the vehicles increases, especially for DCP and LTE4V2X-C without multi-hops. This is a direct repercussion of the effect shown in Figure 4.18, where we have noticed that the packet loss decreases when the speed of the vehicles increases.

4.6.2.2. *LTE4V2X-D*

Figure 4.20 shows the evolution of the delivery rate of the information that needs to be disseminated, depending on time, for LTE4V2X-D and LTE4V2X-D without multi-hop extensions. The delivery rate corresponds to the percentage of the vehicles that are in the notification area (see Figure 4.13) and that have rightfully received the message that needs disseminating. Thus, at the beginning of the simulation, no vehicle is present in the notification area, and the delivery rate is therefore of 100%. At $t = 49$, the delivery rate decreases, because we have one or several vehicles that have entered the area to be notified, but that have not received the message. It therefore takes a certain amount of time (~20 s for LTE4V2X-D without multi-hop, almost instant for LTE4V2X-D) for the entering vehicles to receive the message and for the delivery rate to go high up again. This curve therefore clearly shows that LTE4V2X-D informs the vehicles that enter the notifying zone very quickly because of its multi-hop extension.

4.6.3. *Analysis of the impact of the handover*

In this section, we assess the impact of the handover on LTE4V2X-C: since the FCD application is periodic, we have assessed the delay between two FCD packets when the vehicle moves on to a handover toward an eNodeB adjacent to it. We have calculated this delay for the two protocols LTE4V2X-C and DCP.

When a vehicle arrives in a new cell, the time that it will have to wait before its first FCD transmission depends exclusively on its arrival instant in the new cell in relation to the round of its new eNodeB. The two following sections will detail this point. We do not consider the delay of the LTE handover in itself, because that is the same for the two protocols.

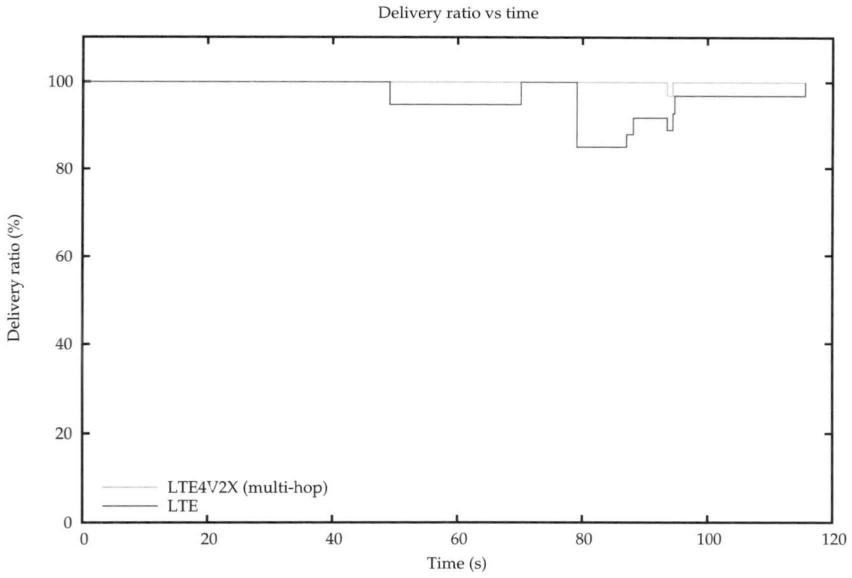

Figure 4.20. *Delivery rate depending on time*

4.6.3.1. *Impact of the handover on the LTE4V2X-C protocol*

When the vehicle is connected to its new eNodeB, it must wait for a cluster update packet on the part of the eNodeB, and then a given time that depends on the identifier that has been attributed to it, so as to avoid the simultaneous communications with the vehicles that could arrive at the same time as itself in the cell. We have identified the best and worst cases, and we have deduced the time that a vehicle must wait before transmitting its first FCD packet to its new eNodeB (see [4.3], where t is the instant that the vehicle has arrived, $t = 0$ corresponding to the end of the configuration phase).

$$\begin{aligned} f(t) &= \text{Duration of a round} + 0.001 + 0.001 \cdot \text{ID} - t \\ &\quad \text{if} \quad 0 \leq t < \text{Duration of a round} \end{aligned} \qquad [4.3]$$

4.6.3.2. *Impact of the handover on the DCP protocol*

The vehicle waits a period of time that is randomly generated between 0 and the duration of a round before sending out its first FCD packet in the VANET. This random parameter allows us to avoid the simultaneous send outs in case several vehicles arrive at the same time. We have considered the best and worst cases, and have deduced the time that a vehicle must wait before transmitting its first FCD packet to its new eNodeB (see [4.4], where t is the instant that the vehicle arrives, $t = 0$ corresponding to the end of the configuration phase).

$$\begin{aligned} g(t) = {} & 1.5 \cdot \text{Duration of a round} \\ & + \text{Duration of the configuration phase} \\ & + \text{Duration depending on the ID of the CH} - t \\ & \text{if} \quad 0 \leq t < \text{Duration of a round} \end{aligned} \qquad [4.4]$$

If we apply these two functions to a round duration of 1 s, a configuration phase of 0.5 s, and an average number of 400 vehicles in the LTE covering area, we obtain for LTE4V2X-C and DCP an average time of, respectively, 0.7 s and 2.202 s knowing that a node does not transmit its first FCD to the new eNodeB. Then, if we suppose that the vehicle has sent its last FCD to its old eNodeB in $t - 0.5$ s, the average time between two FCD packets when a handover takes place is, respectively, 1.200 s and 2.702 s for LTE4V2X-C and DCP.

Figure 4.21 shows, for LTE4V2X-C and DCP, the time that flows between the receipt of the last FCD packet via the eNodeB source, and the reception of the first FCD packet via the destination eNodeB, depending on the instant where the vehicle arrives in the round of its destination eNodeB in relation to the configuration phase. This curve was traced starting from [4.3] and [4.4], with an offset that corresponds to a half-round, in order to reflect the duration between the last FCD transmission to the eNodeB source and the beginning of the handover. We thus observe that DCP induces, in all the cases, more delay than LTE4V2X-C.

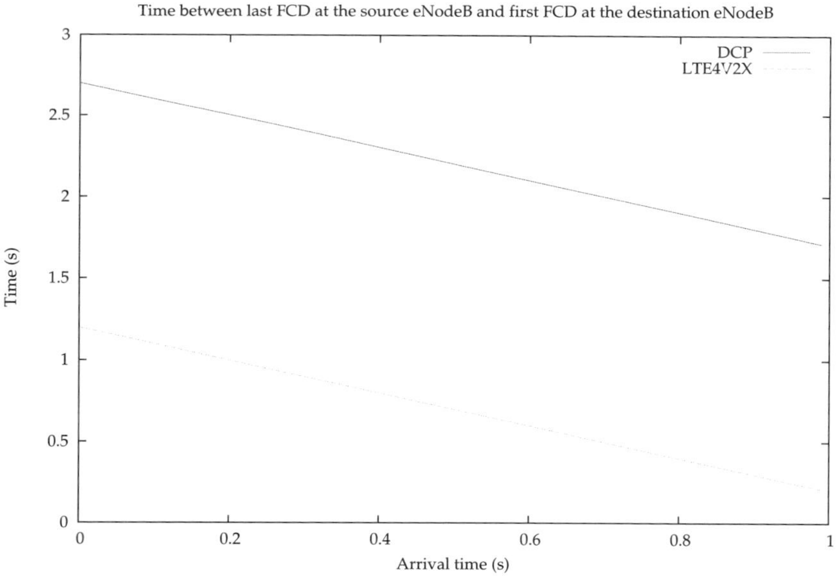

Figure 4.21. *Duration of the handover*

4.7. Conclusion

This chapter presents LTE4V2X, a complete solution that uses the operating network LTE for controlling and managing an VANET in order to improve its performance. The solution proposes a new centralized organization architecture of the VANET. It also considers the case of non-coverage by the LTE network (in the presence of a tunnel, for example), and proposes a localized distributed extension in the non-covered areas. This innovative architecture allows us to make the best use of the LTE and VANETs, making them complementary, and thus optimizing the use of wireless channels in VANETs. This allows for an available place for all the other applications in the vehicle network, whether it is road safety applications, entertaining applications or driving assisting applications. To test the performances of the solution, we have implemented two protocols: a collection protocol LTE4V2X-C and a dissemination protocol LTE4V2X-D. The two protocols have been compared to decentralized approaches. The results show that our collection protocol LTE4V2X-C is better than a decentralized approach, in the sense that it creates less overload in the network and only loses a very small percentage of useful data. The results also show that our dissemination protocol LTE4V2X-D is very efficient for disseminating information in a determined area, even when part of this area does not receive the LTE signal. The extensions that consist of adding the multi-hops in LTE4V2X have greatly improved their performances for the LTE cells where the coverage is not complete. LTE4V2X turns out to be very efficient for an urban topology as well, both at a high and low vehicle density. Our future works will try to extend LTE4V2X to real-time applications, and to study in more detail the effects of the handover on the behavior of our protocols.

4.8. Bibliography

[BAJ 09] BAJABER F., AWAN I., "Centralized dynamic clustering for wireless sensor network", *IEEE Advanced Information Networking and Applications*, Bradford, UK, 2009.

[BEN 11] BENSLIMANE A., TALEB T., SIVARAJ R., "Dynamic clustering-based adaptive mobile gateway management in integrated VANET-3G heterogeneous wireless networks", *IEEE Journal on Selected Areas in Communications*, vol. 29, no. 3, pp. 559–570, march 2011.

[BON 07] BONONI L., FELICE M.D., "A cross layered MAC and clustering scheme for efficient broadcast in VANETs", *International Workshop on Mobile Vehicular Networks (MoVeNet 2007)*, Pisa, Italy, 2007.

[CAR 10] CARDOTE A., SARGENTO S., STEENKISTE P., "On the connection availability between relay nodes in a VANET", *IEEE Global Telecommunications Conference*, Miami, FL, 2010.

[CHA 97] CHAO C., KATURA T., KISIMOTO M., *et al.*, "Network connectivity of layered self-organizing wireless networks", *Proceedings of the 6th International Conference on Computer Communications and Networks 1997*, Las Vegas, NV, 1997.

[CHE 09a] CHERIF M., SENOUCI S.-M., DUCOURTHIAL B., "A new framework of self-organization of vehicular networks", *IEEE Global Information Infrastructure Symposium*, Hammamet, Tunisia, 2009.

[CHE 09b] CHERIF M., SENOUCI S.-M., DUCOURTHIAL B., "Vehicular network self-organizing architectures", *IEEE GCC Conference*, Kuwait, 2009.

[CHI 02] CHISALITA L., SHAHMEHRI N., "A peer-to-peer approach to vehicular communication for the support of traffic safety applications", *IEEE Intelligent Transportation Systems*, Singapore, 2002.

[DUR 10] DURRANI S., ZHOU X., CHANDRA A., "Effect of vehicle mobility on connectivity of vehicular ad hoc networks", *IEEE Vehicular Technology Conference Fall*, Ottawa, Canada, 2010.

[EUR 12] EURECOM, VanetMobiSim website, 2012. Available at http://vanet.eurecom.fr/.

[FOU 12] FOUNDATION N.S., The Planète Group at INRIA, NS-3 website, 2012. Available at http://www.nsnam.org/.

[KAI 11] KAISSER F., JOHEN C., VÈQUE V., "Etude de la formation de convois dans un réseau de véhicules sur autoroute", *Colloque Francophone sur l'Ingénierie des Protocoles (CFIP 2011)*, Hôtel Les Jardins, Sainte-Maxime, France, pp. 559–570, 2011.

[SAL 09] SALHI I., CHERIF M., SENOUCI S.-M., "A new architecture for data collection in vehicular networks", *IEEE International Conference on Communications*, Dresden, Germany, 2009.

[TAL 10] TALEB T., BENSLIMANE A., "Design guidelines for a network architecture integrating VANET with 3G & beyond networks", *IEEE Global Telecommunications Conference*, Miami, FL, 2010.

[WAN 08] WANG S., YANG C., LIN K., "A self-organizing backbone construction scheme for efficient inter-vehicle communications", *International Workshop on Wireless, Ad Hoc, Mesh and Sensor Networks 2008*, Hainan, China, 2008.

[YIZ 10] YIZHI W., JIANMING H., QI W. *et al.*, "A study of distributed traffic information acquisition based on clustered VANET", *IEEE International Conference on Optoelectronics and Image Processing*, Haiko, Hainan, China, 2010.

Chapter 5

Gateway Selection Algorithms in Vehicular Networks

5.1. Introduction

IEEE 802.11-based vehicular *ad hoc* networks (VANETs) have been widespread due to their relevant attractive features such as self-organization and decentralized administration. The great potential of this technology has been acknowledged with the establishment of ambitious research programs on vehicular communication systems worldwide, such as the European eSafety framework, numerous United States vehicle-to-vehicle (V2V) and vehicle-to-infrastructure (V2I) projects, and the Japanese Smartway and Advanced Safety Vehicle programs. Vehicular communication and networking present an active field of standardization activities worldwide, such as ISO TC204, IEEE (802.11p and 1609.x) and SAE DSRC in the United States, ETSI TC ITS and CEN WG278 in Europe and ARIB T-75 in Japan. In addition to the worldwide standardization

---

Chapter written by Ghayet el mouna ZHIOUA, Houda LABIOD, Nabil TABBANE and Sami TABBANE.

efforts, considerable evolution inside the vehicles themselves is observed. In fact, future vehicles are expected to be equipped with high-efficiency computing systems and multiple wireless communication interfaces, that is on board units (OBUs). Thus, intelligent transportation systems (ITS) attract not only the research community but also the automotive industry. Recently, they have focused their efforts on bringing vehicular communication and networking to maturity by moving it from the research field into real implementation, aiming to provide not only more safety in the transportation systems but also other high quality of service (QoS)-based services and applications for their customers. Co-operative wireless communications between *ad hoc* networks and cellular systems emerge as an attractive common paradigm, especially in the ITS domain.

VANET networks are crucial for ITS systems. In fact, future vehicles are expected to be equipped with high-efficiency computing systems and multiple wireless communication interfaces. According to the ETSI 102 638 technical report, in 2017, 20% of the running vehicles will have communication capabilities and by 2027 almost 100% of the vehicles will be equipped with OBUs. OBUs are devices that provide communications among neighboring vehicles, that is V2V or between vehicles and nearby fixed equipment, that is V2I communications. In this chapter, we focus on studying a hybrid network architecture composed of two systems: a VANET network and a fourth-generation (4G) cellular infrastructure, under the context of evaluating the impact of mobile *ad hoc* network (MANET) integration into a cellular system. More precisely, we focus on gateway selection in a hybrid clustered VANET-LTE advanced cellular network architecture. We first present the state of the art of clustering and gateway selection for traditional V2I communication in VANET networks. Then, we present the method used to ensure a co-operative traffic transmission in a joint 4G LTE-advanced and IEEE 802.11p-based clustered VANET network.

5.2. Clustering and gateway selection in VANET networks

VANET networks have to overcome some issues and challenges related to their specific characteristics: which are the very dynamic network topology due to vehicles high velocity, aiming to ensure acceptable V2V and V2I communications. Most of the solutions proposed for handling these issues are based on the creation of dynamic clusters to self-organize the IEEE 802.11p vehicular network. Dynamic clustering formation can be done either be in a decentralized or centralized way. The following sections present the clustering problem statement and the state of the art of clustering algorithms in detail.

5.2.1. *Clustering in VANET networks*

5.2.1.1. *Problem statement*

As in a classic MANET network, VANET nodes have dynamic connectivity and self-organizing features. However, with an increase in the number of nodes, where each node handles its own decentralized routing and neighborhood connectivity maintenance tasks, serious scalability and hidden terminal problems may occur. The most common solution adopted for this problem is clustering. Clustering in VANET aims to organize vehicles into groups based on some specific common characteristics. Using this technique can lead to more node coordination and less internode interference. The idea behind clustering is to group the network nodes into a number of overlapping clusters. Typically, the members of a cluster have some common characteristics, for example close coordinates, velocities and the same direction. In addition, clustering in VANETs requires selecting the cluster head (CH) to be responsible for coordination from amongst the members of its cluster. The selection process of CH is carried out by all nodes belonging to the cluster, where each node broadcasts its information to all other neighboring nodes. After information collection from all the nodes and based on a CH selection algorithm, see section 5.2.2.2, these nodes select a CH. After the decision regarding the CH is completed and the cluster nodes are informed, the CH will be able to communicate directly with all the members of its cluster and coordinate the communication among them, and will act as the relay node (RN) of communications to other cluster members and other nodes in different clusters.

Clustering makes a hierarchical routing, in which paths are recorded between clusters instead of between nodes, possible. This increases the route lifetime, thus decreasing the amount of routing control overheads. Inside the cluster, a node that coordinates the cluster activities is the CH. Inside the cluster, there are also ordinary nodes that have direct access only to this CH, and gateways. Ordinary nodes send the packets to their CH, which either distributes the packets inside the cluster or (if the destination is outside the cluster) forwards them to a gateway node to be delivered to the other clusters or to the infrastructure. By replacing the nodes with clusters, existing routing protocols can be used in the network where only intercluster gateways and CHs participate in the propagation of signaling and routing messages. In dense networks, this significantly reduces the routing overhead, thus solving scalability problems for routing algorithms in large *ad hoc* networks.

In the next section, we list some proposed clustering algorithms in MANETs.

5.2.1.2. *State of the art*

The main challenge for a VANET clustering algorithm is to maintain cluster stability for the longest period; otherwise, the performance will be degraded due to the frequent reclustering operations. Several clustering algorithms have been proposed in the literature. In this section, we present an overview of recently proposed clustering algorithms in MANETs.

In [DRO 11], the authors propose a hierarchical clustering technique where cluster members are grouped into subsets of slave nodes and cluster relay nodes and a CH that is at the top of the hierarchy. This algorithm generates a huge amount of overheads due to the clustering hierarchy maintenance.

The algorithm proposed in [KWO 03] is a clustering protocol that does not use special control packets dedicated to perform clustering, but builds the cluster and maintains it based on data traffic forwarding. This algorithm is suitable for a dense network with high mobility because the cluster maintenance depends on the traffic. Moreover, it is not affected by the increase in the control overhead caused by the frequent changes of cluster members. However, this algorithm does not consider the relative velocity metric that causes a decrease in the lifetime of the cluster.

The clustering method proposed in [ELG 08] is built upon the protocol of [KWO 03] and extends the lifetime of the network by balancing energy consumption among the network nodes.

Basu *et al.* [BAS 01] propose the MOBIC algorithm, which uses a special mobility metric for the cluster formation phase where each mobile node sends two consecutive messages to each of its neighbors to compute their relative speed. Then, each mobile node broadcasts this information to its neighbors. The drawback of this method is the need for extra explicit message exchanges among mobile nodes for maintaining the cluster structure. Thus, with frequent network topology changes resulting in frequent clustering updates, cluster maintenance overhead would increase drastically, consuming a high portion of the bandwidth

In [KUK 09], the authors propose a density-based clustering algorithm that takes into account the effects of multipath fading. The cluster formation is based on the weight metric that takes into consideration the link quality and the traffic conditions.

A position-based clustering technique is proposed in [WAN 08], where the cluster structure is determined by the geographic position of the nodes. The stability of the system is improved by selecting the vehicles that have a longer trip as cluster heads. Despite it seeming that this solution gives stable clusters, performance simulations and evaluation do not consider sparse and jammed traffic conditions that are very frequent in the VANET environment.

A similar approach is defined in [JER 07], where clusters are formed on the basis of the vehicles' position on the road. However, the proposed algorithm is very limited as it does not address the cluster maintenance and the CH selection challenge.

Another position-based clustering algorithm that performs hierarchical and geographical data collection and dissemination is proposed in [SAL 09]. The cluster formation in this algorithm is based on the division of the road into segments. Its performance is affected by the mandatory availability of an infrastructure. Moreover, it generates high overheads for V2V and V2I communications.

In [FAN 08], the authors describe a theoretical analysis of the directional-based clustering algorithm. Cluster formations are based on three mobility metrics: moving direction, projected distance variation of all the neighboring nodes over time and leadership duration. The stability of the cluster with fewer overheads is demonstrated.

In [MAS 09], the cluster formation is based on the direction of the vehicles, where vehicles moving in the same direction belong to the same cluster.

The authors of [BEN 11] propose a clustering algorithm for a heterogeneous network based on vehicular and Universal Mobile Telecommunications System (UMTS) cellular networks. The clustering is based on three criteria: the direction of movement, UMTS received signal strength (RSS) and 802.11p wireless transmission range. Neighboring vehicles with the same direction of movement and an RSS higher than a specific threshold belong to the same cluster. The drawback of this approach is that one cluster could be composed of a large number of nodes, which generates a huge amount of overheads for cluster maintenance.

5.2.2. *Gateway selection in a clustered/non-clustered VANET architecture*

Under the scope of V2I communications, an initial set of services for use by V2I systems on European highways has been defined within the European Commission project on co-operative systems for intelligent road safety (COOPERS) [BEL 10]. They defined safety-critical services, such as accident warning, and roadwork information and convenience services, such as journey time and road charging services. Therefore, it is essential to achieve good V2I communications to ensure the proper use of VANET networks. The next sections present in detail the problem statement of the selection of the gateway to the infrastructure and its corresponding related works.

5.2.2.1. *Problem statement*

In a clustered architecture, in spite of its role of initiating the communication and controlling the flow of signaling messages within the cluster, the CH is also set as the gateway to the infrastructure for the source vehicles of its cluster. Obviously, centralizing the connectivity to the infrastructure for nodes belonging to the same cluster is beneficial. The key advantage is to decrease the cellular network resource consumption, by multiplexing distinct source nodes' flows into one gateway that handle sending them to the infrastructure. However, this scheme generates serious issues such as causing CH overload, increasing end-to-end delay as compared to the direct link (that is send directly to the eNodeB) that is intolerable for delay-sensitive services. Moreover, the CH might not be the optimal gateway to the infrastructure as almost all algorithms consider only VANET layer features for CH selection and forget reflection about the infrastructure layer features such as base station load and RSS. References that do not consider *ad hoc* node clustering select the gateway after information broadcast and gathering from all the nodes leading to a huge amount of overheads in the network.

5.2.2.2. *Related works*

In this section, we present several algorithms that have been proposed for CH and gateway selection to the infrastructure in clustered and non-clustered *ad hoc* architectures.

For a non-clustered *ad hoc* network, in [FU 08], the authors proposed an algorithm for gateway selection based on the choice of the mobile node with shortest hops from the source node. One single metric that combines physical hops and virtual hops relative to congestion and contention levels is used. This algorithm uses the network allocation vector (NAV) timer to compute virtual hops for contention measurements, which is not an easy parameter to evaluate concretely.

In [MAN 09], the authors discuss the issues associated with the selection of mobile gateways in an integrated MANET-UMTS heterogeneous network. They use simple additive weighting techniques to select an adequate gateway based on residual energy, UMTS signal strength and mobility speed of the gateway candidates.

In [SET 08], three metrics are used to select the gateway that interconnects MANET with the infrastructure network: energy, mobility and the number of hops based on a simple additive weighting method. The node with the highest weight will be selected as the gateway. These three metrics are not enough to select the optimal gateway as the link from the gateway to the infrastructure is not considered.

In [LIA 10], the authors propose dispersion sorting mechanism (DSM). With DSM, the coverage of the road side unit (RSU) is divided into two areas. One is an area that has a good quality of communication, and the other is an area that has poor quality of communication. The selected gateway, called the RN, collects data from vehicles and uploads data to the RSU. The advantages of this method are that it expands the coverage of the RSU's service and reduces the window period between RN and RSU communication. The gateway selection criteria are as follows: first achieve a good quality of communication with RSU and second a longer time of communication with RSU. The authors formulate these criteria using the positions and velocity of vehicles.

In [AMD 09], the authors propose a Scan-based AP Selection (SAPS) and History-based AP Selection (HAPS) method for infrastructure selection, called access point (AP) selection. Instead of selecting a single AP as in traditional schemes, SAPS sorts the scanned APs and generates an ordered list that is destined to ensure a solid connectivity to the vehicle. Each time the vehicle is disassociated from an AP, it picks out the next AP in the list. With HAPS, scan operations are not required. It exploits road maps built by the vehicle from previous drives or via an exchange protocol between vehicles for the purpose of reducing delay overheads by eliminating scan operations.

For clustered networks, clustering algorithms are used in VANET networks to ensure stability and increase link lifetime between vehicles belonging to the same cluster. There are basic clustering and CH selection techniques also proposed in the literature such as the highest degree [RAM 02] and the lowest Id [BAS 99] algorithms that are not efficient enough as they may generate frequent reclustering.

In [MAS 09], considering that cluster formation is based on direction of vehicles, the first vehicle moving in that direction will be selected as the CH. Using this method, vehicles with a relatively high speed will generate frequent CH reselection that causes additional overheads in the cluster.

In [SAL 09], the authors propose to select one CH in each segment of a road based on geographical information collection which are provided by the infrastructure. This algorithm fails to address cluster stability and cluster maintenance

In [WAN 08], vehicles having a longer trip are selected as CHs. This selection process is based on the priority associated with each node. The priority of the node is calculated using a hash function based on the estimated travel time of the node.

In [ALM 10], the CH is selected on the basis of an additive metric of three criteria: network connectivity level (based on the maximum number of vehicles that are directly connected to considered vehicle and on the vehicles on the same traffic flow) and average distance and velocity levels. The authors consider urban scenarios characterized by several lanes and the intersection architecture, for example roads going straight through or turning left or right. Gathering information over the roads' topology and flow intersections requires accurate positioning systems, which is not always achieved. Moreover, the high dynamicity of the vehicular nodes and the random drivers' reactivity causes limits to this approach.

In [BAS 01], the authors select the CH based on relative velocity. The relative velocity is calculated using a ratio of RSS of two successive HELLO messages. The CH is then the node with the lowest relative velocity variance. This is an interesting approach for selecting the CH; however, there are more accurate techniques for calculating the relative velocity than using the HELLO message-based approach.

In [DRO 11], after discussing the hierarchical clustering technique, the authors propose to select the CH as the slave node that receives three synchronous messages. This method does not consider the vehicles' movement dynamicity and causes a huge amount of overheads to the messages exchanged for the hierarchical cluster maintenance.

In [BEN 11], the CH is designed as the vehicular node that is in the middle of the cluster, at equal distance from the border nodes. A source vehicle will then select one of the selected CHs as gateway to the infrastructure.

In [LUO 10], the geographical area is divided into a series of logical grids and each of them will have an exclusive ID. A cluster header is selected in each grid.

5.2.3. *Conclusions*

The idea behind clustering is to devise the VANET network into a number of overlapping clusters, where each cluster is composed of one or several vehicular nodes. Typically, the goal of the clustering algorithm is to form clusters where members of one particular cluster possess some common characteristics, for example close coordinates, velocities and the same direction, to ensure high V2V link connectivity duration. Moreover, due to the safety-critical services that it handles, it is essential to have good V2I communications by selecting the best gateway to the infrastructure that presents good VANET and V2I features.

5.3. Gateway selection in a clustered VANET-LTE advanced hybrid network

In this section, we describe a hybrid network architecture composed of two systems: a clustered VANET network and an LTE-advanced infrastructure (as a 4G cellular network), under the context of evaluating the impact of VANET integration into cellular systems. In fact, in a clustered topology, the CH is usually set as the default gateway to the infrastructure for all source vehicles of the cluster. Although this architecture is beneficial, it has significant drawbacks, especially in the case of a high-loaded CH. In this section, we address this challenge and present a QoS-balancing gateway selection algorithm (QGwS) aiming to select the appropriate gateway for a given source vehicle under the scope of V2I communications and in a clustered VANET architecture.

5.3.1. *Problem statement*

In a clustered architecture, in spite of its role of initiating the communication and controlling the flow of signaling messages within the cluster, the CH is also set as the gateway to the infrastructure for source vehicles of the cluster. Obviously, centralizing the connectivity to the infrastructure for nodes belonging to the same cluster is beneficial. The key advantage is to decrease the cellular network resource consumption by multiplexing distinct source node flows into one gateway that handles sending them to the infrastructure. However, this method generates serious issues such as causing CH overload, increasing end-to-end delay as compared to the direct link (that is sending directly to the eNodeB) which is intolerable for delay-sensitive services. Moreover, the CH might not be the optimal gateway to the infrastructure as almost all algorithms consider only VANET network features for CH selection and forget reflection about the infrastructure features such as base station load and RSS. Therefore, it is important to propose a method that considers traffic priority and sensitivity on one side and becomes aware of VANET and infrastructure features on the other side. Moreover, to the best of our knowledge, there is no proposed algorithm for gateway selection to infrastructure that considers traffic priority as a criterion. We then propose a multi-criteria and QoS-related attributes approach used to make a decision of the appropriate gateway that might be used to connect a source vehicle to the LTE-advanced infrastructure.

5.3.2. *LTE-advanced standard*

3GPP Release 10 standard specification has been lately released. 3GPP Release 10, also called LTE-advanced, is a well-designed system that meets the requirement of 4G cellular systems. LTE-advanced is a well-designed and very advanced system developed to be adapted with the existing 3G cellular architecture. This is important from an operator point of view; thus, it allows the soft integration of a new system without facing the high costs of a new deployment and dropping the existing investments. LTE-advanced introduction is handled just by software upgrades for most of the LTE Release 8 entities. It is also important from a user point of view that new proposed systems achieve backward compatibilities with the already acquired user's terminal. LTE-advanced, also called 3GPP Release 10, is based on LTE Release 8/9 with adding some enhancements. The first release of LTE, LTE Release 8, provides only a UL peak rate up to 75 Mbit/s, a DL peak rate up to 300 Mbit/s and 4 bit/s/Hz for UL peak spectrum efficiency. However, it presents a delay of about 5 ms and a DL peak spectrum efficiency of 16 bit/s/Hz, which meets the IMT advanced requirements. Besides, LTE supports spatial multiplexing and transmission diversity achieved by using MIMO antennas. Beamforming is also discussed. The LTE evolved packet system (EPS) consists of the core network which is the evolved packet core (EPC) and the evolved UTRAN (E-UTRAN) which contains only the E-UTRAN NodeB (eNodeB); see Figure 5.1. The eNodeB represents the radio access network. It provides radio access user plane (PHY/MAC/RLC/PDCP) and control plane (RRC) protocols toward the user equipment (UE). The EPC is composed of four entities, a mobility management entity (MME) serving GW + PDN GW (SAE-GW) and a policy charging rules function (PCRF)), and provides IP connectivity between a UE and an external packet data network (PDN).

With LTE Release 9, which was released at the end of 2009, only a few enhancements were added to the previous release such as support for broadcast and multicast services (E-MBMS), defining new self-optimization features for self-organizing network (SON) capabilities, etc. SON functions can be divided into two categories: self-configuration and self-optimization. The self-configuration task is a set of functions that enables the network to automatically perform an installation procedure, that is the plug and play, and the self-optimization task is a set of functions that enables the network to auto-tune its operational parameters using UE, eNodeB and performance measurements.

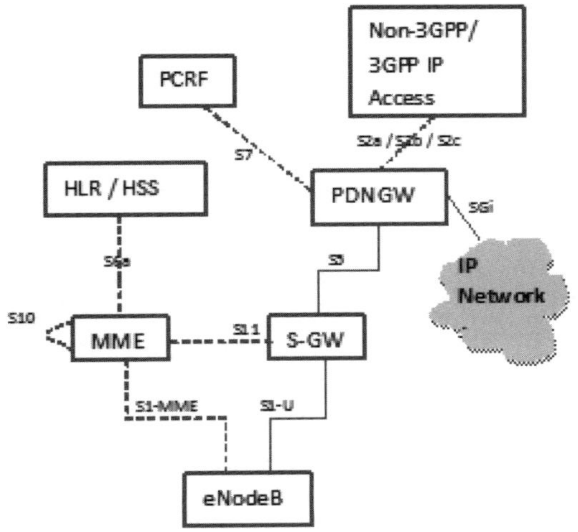

Figure 5.1. *LTE-advanced architecture*

LTE Release 10 will probably not have a new system architecture but it will be evolution of LTE Release 8/9 to further improve the network performance to meet IMT-Advanced requirements. The new added features for LTE Release 10 are carrier and bandwidth aggregation, enhanced MIMO, co-operative multi-site transmissions and relaying. The paired spectrum used is no longer symmetric due to the traffic asymmetry between DL and UL. Moreover, with LTE Release 10, transmission bandwidth is extended up to 100 MHz to satisfy the requirement for the peak data rate. In fact, a wider-band transmission is achieved by aggregating multiple carriers for only one joint transmission to and/or from one UE; this is called carrier aggregation (CA). The main advantages of the CA for the operators are the ability of offering high data rate services by aggregating non-adjacent component carriers with different bandwidths. Operators are no longer limited by having a single wideband spectrum bandwidth to achieve high data rates. In addition, in the UL, improvements are necessary to satisfy the 6.75 bit/s/Hz of UL peak spectrum efficiency set by the IMT-Advanced requirement. LTE Release 10 holds this issue by the enhancements of MIMO techniques with introducing the single-user MIMO with up to four transmissions. In the same way, in the DL, although IMT-Advanced requirement for peak spectrum efficiency is satisfied, further improvements are necessary while considering the traffic increasing demand and cell-edge user. LTE Release 10 holds this issue by extending the number of transmissions from a maximum of four in LTE Release 8 to a maximum of eight in LTE-advanced. In addition, co-operative base station capabilities are possible due to the X2 interface. eNodeBs can coordinate radio resources scheduling to decrease intercell interferences. Signal strengthening is possible using eNodeBs joint transmission and coordinated multi-site beamforming [GPP 11]. Besides, as coverage problems increase, due to the use of high-frequency spectrum and as the capacity improvement could only be achieved by small cell sizes, LTE-advanced has introduced relaying in E-UTRAN as a solution for these problems. It is called the multihop architecture because it combines the benefits of both *ad hoc* and cellular networks. LTE-advanced multihop architecture is also used to increase the coverage, especially at cell borders, and capacity by reducing the transmitting distance through multihop links which will achieve higher data rates as compared to the long single links. With relaying, the UE communicates with the network via an RN that is wirelessly connected to a donor cell using the LTE radio interface technology.

3GPP Release 10 is a very advanced and well-designed system for 4G cellular networks. Due to its relevant features and as it achieves backward compatibility with 3G systems, LTE-advanced standard might be widely adapted by numerous operators as the next generation of their cellular networks.

5.3.3. *Proposed algorithm*

We propose in this section a multi-criteria and QoS-related attribute approach that aims to make a decision of the appropriate gateway that might be used to connect a source vehicle to the LTE-advanced infrastructure.

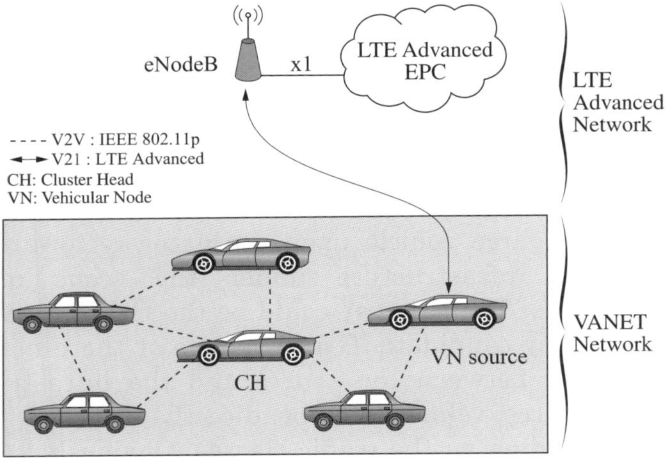

Figure 5.2. *System model*

5.3.3.1. *System model*

The system model is based on a hybrid network architecture that consists of two systems: an LTE-advanced infrastructure and a VANET network. The architecture comprises IEEE 802.11p-based VANET vehicles and an LTE-advanced eNodeB connected via an S1 interface to the core network, that is the EPC; see Figure 5.2. We assume that all vehicles are equipped with OBUs that contain two communication interfaces: an IEEE 802.11p interface and an LTE-advanced interface. Each vehicle uses its VANET interface to communicate with its neighboring vehicles based on the IEEE 802.11p standard. IEEE 802.11p intervehicle communications are used even if they are inside the base station coverage. Note here that the simultaneous two interface communications do not cause harmful interferences as they lie on distinct non-overlapping channels in two different spectrum regions.

We made the assumption that the VANET network is already clustered and the CH is selected. Clustering is used in VANET networks to overcome intervehicle connectivity problems caused by high mobility and relative velocity of the vehicles by dividing the vehicular network into homogeneous sets. Clustering and CH selection techniques are out of the scope of this chapter. The goal of this chapter is to make the decision of a source vehicle upon the gateway towards the LTE-advanced infrastructure under the scope of V2I communications. The gateway could be the CH of its cluster, another gateway candidate (GWC) vehicle or there may also be no gateway between the source and the infrastructure where the source vehicle will be directly attached to the eNodeB, depending on the traffic classes. Obviously, sending the traffic of each source vehicle directly to the infrastructure without defining a gateway is in theory the ideal scenario or method. However, nowadays, cellular networks still suffer from a lack of resources in spite of numerous propositions for the optimization of radio resource allocation schemes. Thus, decreasing cellular network resource consumption, by multiplexing distinct source nodes' flows into one default gateway, that is the CH, which manages to send them to the infrastructure, seems to be an advantageous solution for a clustered VANET architecture. Unfortunately, although it overcomes the lack of radio resources, this method generates serious issues such as causing CH overload and increasing end-to-end delay, which is intolerable for delay-sensitive, real-time and alert services, etc. Therefore, the key challenge we address in this chapter is how to select for a source vehicle the appropriate V2I gateway while considering at the same level the QoS traffic class, the CH features and the cellular network load. The next section presents our proposed algorithm which is optimized by the integration of fuzzy logic.

5.3.3.2. *QoS-balancing gateway selection algorithm*

An efficient gateway selection algorithm is one that considers QoS features of traffic to be transmitted to the infrastructure. We have designed a QGwS in a VANET-LTE-advanced hybrid network where the decision of the gateway depends on the traffic to be transmitted to the infrastructure [ZHI 12]. It is an adaptive multi-criteria multi-attribute gateway selection decision algorithm. In this section, we propose an improvement of our algorithm by adding the fuzzy logic for criteria and attributes check because of the inherent strength of fuzzy logic in solving problems exhibiting imprecision and the fact that many of the terms used for describing radio signals are fuzzy in nature [BCH 10]. This is the fuzzy QoS-based gateway selection (FQGwS) algorithm. Fuzzy logic can be viewed as a theory for dealing with uncertainty about a complex system, and as an approximation theory. Fuzzy logic has two objectives: first, it develops computational methods that can perform reasoning and problem-solving tasks that require human intelligence and, second, it explores an effective trade-off between precision and the cost in developing an approximate model of a complex system [BCH 10]. A fuzzy inference system is a computing framework based on the concepts of fuzzy set theory, fuzzy if-then rules and fuzzy reasoning. Inference is known as the process that draws conclusions from a set of facts using a collection of rules. In this chapter, we use the Mamdani fuzzy inference system [JAN 95]. The Mamdani fuzzy system is composed of modules: the fuzzifier, the fuzzy rules base, the fuzzy inference engine and the defuzzifier.

The FQGwS algorithm considers QoS requirements of the source vehicle and cluster state. We propose to implement this algorithm in the vehicles having OBUs. The FQGwS uses the following input parameters for making a final decision upon the gateway for V2I communications.

5.3.3.2.1. Input parameters

FQGwS algorithm input parameters are divided into two categories: attributes and criteria.

Attributes

In this section, we consider only one attribute, that is the type of traffic to be transmitted by a source vehicle to the infrastructure. We consider three classes of traffic class 1, class 2 and class 3. As the source vehicle is a VANET node, generated traffic types are governed by the IEEE 802.11p standard. On the other hand, the infrastructure is an LTE-advanced cellular network. Thus, to achieve system compatibility, a mapping between IEEE 802.11p access categories and LTE-advanced priority classes is done. The mapping is performed on the basis of QoS constraints (packet average delay budget and packet loss) of VANET and LTE-advanced standards. Traffic mapping is presented in Table 5.1. We define only three categories of QoS classes because for first deployments the majority of 4G operators will likely start with three basic service classes containing voice, and best effort data classes. In the future, dedicated bearers offering premium services such as high-quality conversational video can be introduced into the network. Moreover, for the sake of simplicity, we will call class 1 traffic voice traffic, class 2 streaming and class 3 data traffic.

QoS Classes	Class 1 (voice)	Class 2 (streaming)	Class 3 (data)
IEEE 802.11p Standard Designation	Voice access category	Video access category	Best effort and background access category
LTE-advanced Standard Designation	Conversational services and control signaling	Streaming and interactive services	Background services
Example of services	Conversational voice, live streaming, Real-time gaming, IMS signaling, safety and alerts services	Conversational video, live streaming	Non-conversation video (buffered streaming) TCP-based, FTP P2P file sharing, e-mail, Web browsing, chat

Table 5.1. *Traffic mapping*

In addition to the delivery of information and entertainment, that is infotainment services, the role of typical V2I communications includes the provisioning of safety-related, real-time and alert services, such as speed limit information, traffic jams, safe distance and accident warnings [BEL 10]. All these services aim to prevent accidents by directly providing information to the drivers of the vehicles. Moreover, ITS protocols are designed for the 5.850- to 5.925-GHz band, divided into one central control channel (CCH) and six service channels (SCHs) where CCH is dedicated to the transmission of traffic safety messages, whereas SCHs are dedicated to the transfer of various application data. IEEE 802.11p standard allows the use of the four traffic classes within the node for each channel

(CCH or SCH) as the actual combination of traffic classes of IEEE 802.11p Medium Access Control (MAC) layer is based on the IEEE 802.11e enhanced distributed channel access (EDCA) function [MIS 11]. We propose to classify traffic safety applications as real-time systems. However, communicating real-time messages requires a predictable system that is able to deliver the message before the deadline. Therefore, inside class 1 traffic, safety and alerts services have priority over other conversational voice or live streaming services. In this section, we consider traffic priority class for choosing the adequate gateway to the infrastructure under the V2I context.

Criteria

Criteria are input parameters measured and received by the source vehicle and used to make the final decision of the gateway to the infrastructure. The criteria of the FQGwS algorithm are nearly the same as those of the QGwS algorithm [ZHI 12]:

– ***Source Vehicle to LTE-advanced infrastructure link connectivity strength (Source2I LCS)***: as the source mobile vehicle could detect one or several eNodeBs with various loads, we define a simple additive metric (SAM) of two important metrics: the RSS and the load (L) to select the best base station. In fact, the source vehicle builds the diversity set containing received BSs and selects the eNodeB with the highest LCS_{S2I} computed using [5.1]. The load describes the resource occupation ratio of the eNodeB. This is one of the most important criteria for base station selection in cellular networks nowadays.

LCS_{X2I} between X and the LTE-advanced infrastructure is computed using [5.1], where $X \in \{Source,\ CH,\ GuC\}$:

$$LCS_{X2I} = \max_{i \in [1,n]} \left\{ \frac{RSS_i}{Thr_{RSS}} + \frac{Thr_{Load}}{L_i} \right\} \qquad [5.1]$$

where n is the number of eNodeBs in the diversity set, RSS_i is the RSS of $eNodeB_i$, L_i is the load of $eNodeB_i$ and Thr_{RSS} (= 41.76 dBm) and Thr_{Load} are thresholds of RSS and load criterion, respectively.

It is sometimes difficult for an operator to provide exact information (in %) about its base station load. Therefore, if this is the case, without loss of generality of the LCS criteria, the eNodeB would provide the source vehicle with information about its load by using the set {High, Low, Medium}. A mapping is thus done between this set and its corresponding values {100, 10, 50}, respectively, to compute LCS criteria.

Therefore, the source vehicle selects the best eNodeB from the diversity set ensuring the best compromise between BS RSS and load.

– ***CH to LTE-advanced infrastructure link connectivity strength (CH2I LCS)***: this is the LCS between the CH and the selected eNodeB. The CH selects the eNodeB with the highest LCS_{CH2I} computed using [5.1], where $X = CH$.

– ***Load of the CH***: this is the occupation ratio of the buffering queue of the CH.

– ***Source vehicle to CH link connectivity duration (Source2CH LCD)***: this criterion is used to reflect the stability of the link and its lifetime between two vehicles. In this case, it is the source vehicle and the CH. LCD_{S2CH} is computed on the basis of [5.2]. We cannot consider only relative velocity to characterize link stability between two mobile nodes as we have introduced QoS class characterization. Thus, the link connectivity duration criterion introduces tangibility in V2V link characterization.

LCD_{ij} between vehicle i and vehicle j is computed using [5.2], inspired from [AND 10]:

$$LCD_{ij} = \frac{\sqrt{(\alpha^2 + \gamma^2)R^2 - (\alpha\delta - \beta\gamma)^2} - (\alpha\beta + \gamma\delta)}{\alpha^2 + \gamma^2} \qquad [5.2]$$

where:

$$\alpha = v_i cos\theta_i - v_j cos\theta_j$$
$$\beta = x_i - x_j$$
$$\gamma = v_i sin\theta_i - v_j sin\theta_j$$
$$\delta = y_i - y_j$$

(x_i, y_i), respectively (x_j, y_j), is the Cartesian coordinates of two neighboring vehicles i and j and that have an inclination of θ_i, respectively θ_j(0<θ_i, θ_j<2Π) with respect to the x-axis and moving at v_i, respectively v_j, speed. R is the IEEE 802.11p wireless transmission range.

LCD_{S2CH} is computed on the basis of [5.2]. We suppose that the CH and the source vehicle are adjacent; otherwise, $LCD_{S2CH} = \min\{LCD_{ij}\}$, $1 \leq i, j \leq n$, where n is the number of hops between the source and the CH.

– ***Gateway candidate to LTE-advanced infrastructure link connectivity strength (GwC2I LCS)***: there could be another gateway candidate (GwC) for the source vehicle with better LCS, load and LCD as compared to the CH. Therefore, each ordinary vehicle (OV) that experiences high RSS (>RSS_{th}), low buffering queue load (<L_{th}) and high LCD (>LCD_{th}) becomes a potential GwC (pGwC); see sections 5.3 and 5.4. The latter will select and then keep track of the best GwC, and use it as an input for the FQGwS algorithm for making the final gateway decision. For the GwC2I LCS criterion, the gateway candidate selects the eNodeB with the highest LCS_{GwC2I} computed using [5.1].

– ***Load of the gateway candidate***: this is the occupation ratio of the buffering queue of the gateway candidate.

– ***Source vehicle to gateway candidate link connectivity duration (Source2GwC LCD)***: this is the LCD between the source vehicle and the gateway candidate. LCD_{S2GwC} is computed on the basis of [5.2].

5.3.3.2.2. Proposed algorithm

In this section, we present our proposed algorithm for gateway decision selection, under the scope of V2I communications. FQGwS algorithm is a decentralized scheme performed by a source vehicle that has to connect to the LTE-advanced infrastructure. It is a QoS-balancing scheme as gateway selection is based on the traffic class to transmit to the infrastructure. FQGwS is composed of three phases: the data collection phase, decision phase and maintenance phase.

Data collection phase

As mentioned before, the algorithm input parameters (criteria) are provided using network measurements handled either by the source vehicle and/or participant vehicles (for example CH and OVs). Afterward, the gateway decision is performed.

The first step is the data collection phase, where the dual interface source vehicle triggers measurements on its LTE-advanced and IEEE 802.11p network cards. On the LTE-advanced interface, the source vehicle measures the RSS of neighboring base stations and checks their loads. It then builds the diversity set which contains all BSs it senses and sorts them based on the LCS metric computed using [5.1]. Afterward, the source vehicle selects the eNodeB with the highest $Source2I\ LCS$ to ensure a compromise between RSS and load.

On the IEEE 802.11p interface, as we focus on a VANET clustered architecture, source vehicle is already aware of the CH features, position and velocity. Thus, it deduces $Source2CH\ LCD$. For the remaining criteria, CH sends to the source vehicle its $CH2I\ LCS$ and load.

Decision phase

The second phase is the decision phase where the gateway to the infrastructure has to be selected. In this phase, if the $Source2I\ LCS$ criterion is acceptable and the traffic type is voice, the source vehicle will be attached directly to the best BS without defining a gateway to the infrastructure. This decision is taken because class 1 traffic is very delay-sensitive traffic and not greedy in resources. The source vehicle will check CH criteria for class 1 traffic only if the $Source2I\ LCS$ criterion is not acceptable. However, with the remaining traffic classes, in spite of the $Source2I\ LCS$ criterion, the source vehicle will always check the CH features.

1) Cluster head criteria check

With class 2 and class 3 traffic, the source vehicle verifies the metrics of the default gateway to the infrastructure, that is the CH. In fact, if the CH metrics values, that is $CH2I\ LCS$, CH load and $Source2CH\ LCD$, are acceptable (as compared to the specified threshold of Table 5.2), the decision regarding the gateway is immediately taken without the solicitation of other gateway candidates. The gateway to the infrastructure is then the CH. In fact, the CH has acceptable features when the membership function of each of the three criteria is equal to one. In other words, if $LCS_{CH2I} > LCS_{th}$, CH load $< L_{th}$ and $LCD_{S2CH} > LCD_{th}$ relative to the CH threshold features of Table 5.2.

2) Gateway candidate solicitation

If the levels of the three criteria of the CH are not acceptable (as compared to the specified threshold of Table 5.2), the source vehicle broadcasts a GATEWAY SOLICITATION REQuest message through the cluster to keep track of potential OVs that may become gateway candidates. In fact, each OV with acceptable criteria responds to a GATEWAY SOLICITATION RESponse, containing its metrics features ($GwC2I$ LCS, load, velocity and position (used to compute LCD by source vehicle)). They become potential GwCs (pGwCs). In other words, pGwCs are OVs with high LCS_{GwC2I} ($>LCS_{th}$) and low load ($<L_{th}$) relative to the pGwC threshold features of Table 5.2. Afterward, the source vehicle collects information from all pGwCs to make a decision over the gateway candidate to select by means of weight computation which is based on an additive metric.

QoS Classes		**Class 1**	**Class 2**	**Class 3**
CH	LCS_{th} (RSS_{th}/L_{th})	−90 dBm / 70%	−90 dBm/65%	−90 dBm/75%
	L_{th}	70%	65%	75%
	LCD_{th}	50 s	80 s	80 s
pGwC	LCS_{th} (RSS_{th}/L_{th})	−120 dBm/90%	−120 dBm/85%	−120 dBm/95%
	L_{th}	90%	85%	95%
	LCD_{th}	30 s	60 s	60 s

Table 5.2. *FQGwS algorithm parameters*

3) Gateway candidate selection

Based on the collection criteria of all pGwCs, the source vehicle computes and affects a weight for each pGwC. Weight computations are performed using [5.3], which is a simple additive metric of the three criteria. The criteria are divided into two categories: positive and negative. A positive criterion is a criterion that is considered better while increasing, such as LCS and LCD. However, a negative criterion is a criterion that gets worse as it increases, such as the load. Therefore, the weight W_i of the potential gateway candidate i ($pGWC_i$) can be formulated such as in [5.3].

$$W_i = \sum_j \frac{M_{ij}}{M_{jth}} + \sum_k \frac{M_{jth}}{M_{ik}} \qquad [5.3]$$

where M_{ij} is the positive criterion and M_{ik} is the negative criterion. LCD_{th} and L_{th} are defined in Table 5.2. LCS_{th} is -2.17 for the three traffic classes.

The source vehicle keeps track of the best pGwC with the highest weight, called the gateway candidate (GwC). $GwC2I\ \ LCS$, load and $Source2GwC\ \ LCD$ of the GwC are then used to compute the gateway decision as the input of the FQGwS algorithm.

This algorithm is the baseline for making the decision of the gateway. As can be deduced, a source vehicle checks the CH state in order of priority ($CH2I\ \ LCS$, load and $Source2CH\ \ LCD$). If the latter ensures efficient proxy capabilities, it will be selected as the gateway and the source vehicle will not send the GATEWAY SOLICITATION REQuest message. Otherwise, the source vehicle will solicit other pGwCs. Because of this step, the gateway efficient decision is taken without leading to useless extra overheads due to the GATEWAY SOLICITATION REQuest and GATEWAY SOLICITATION RESponse messages.

Maintenance phase

After making the decision regarding the gateway, the source vehicle initiates its timer and starts transmitting. If there is no gateway between the source vehicle and the infrastructure, the link maintenance phase is handled by periodic channel quality indicator (CQI) reports. However, if a gateway is selected, the transmitting period will last $LCD_{S2CH} - \varepsilon$ or $LCD_{S2GwC} - \varepsilon$ depending on the selected gateway; afterward, if there are more packets to be transmitted, the source vehicle will repeat the process in the CH check phase to reselect the appropriate gateway for the remaining packets.

5.3.3.3. *Performances evaluation*

Through simulations, we compare our FQGwS algorithm based on fuzzy logic modeling with the deterministic gateway selection scheme for clustered VANET architecture, using the CH as the gateway to the infrastructure, such as in [BEN 11, MAS 09]. We present simulation results to confirm the validity of our proposed approach.

FQGwS algorithm attributes, criteria and rules are implemented in Matlab. More precisely, these parameters are LCS, $Load$ and LCD membership functions of each of the three traffic classes for the CH, GwC and $Source\ Vehicle$ and the $FGwD$ output membership function. The rule base is then built on the basis of fuzzy sets of membership functions for the three traffic classes. Simulations were made using Network Simulator NS2.33. We implement an IEEE 802.11p package in order to enable VANET communications among high-speed vehicles and the LTE module for 4G cellular communications. Vehicle mobility is generated with VanetMobiSim [VAN].

5.3.3.3.1. Simulation scenario

In the simulation scenario, we consider a section of a straight multiple-lane highway, where vehicles are moving in the road and are equipped with a Global Positioning System (GPS) receiver. Thus, we assume that each vehicle knows its own position and velocity. The VANET network is organized into clusters where a CH is selected at the top of each cluster. A cluster is composed of N_v vehicles. Clustering and CH selection schemes are out of the scope of this chapter. Hence, we consider two clustering and CH selection algorithms [MAS 09, BEN 11] recently proposed in the literature for VANET networks to evaluate the performance of our FQGwS algorithm. In [MAS 09], the authors propose the C-DRIVE algorithm for cluster formation that is based on the direction of vehicles, where vehicles moving in the same direction belong to the same cluster. The first vehicle moving in each direction will be selected as the CH. This algorithm considers intersection scenarios, while in this section we focus on the generalization of the C-DRIVE to use it as a comparison base for the CH selection algorithms. In [BEN 11], the CH is the vehicle in the middle of the cluster at equal distance, in terms of the number of hops, to the border vehicles of the cluster, called here Middle algorithm. On the other hand, the infrastructure is composed of one eNodeB positioned at the border of the highway. Each vehicle is attached to the eNodeB using its LTE-advanced interface.

In the traffic simulation scenario, a source node that belongs to the cluster uses the FQGwS or the deterministic scheme for gateway selection. In the deterministic scheme, that is C-DRIVE or Middle algorithm without FQGwS, the selected gateway is always the CH. With the FQGwS algorithm, the source vehicle makes the decision of the gateway based on computed input criteria. C-DRIVE and Middle algorithm without FQGwS algorithms are compared to C-DRIVE and Middle algorithm with FQGwS schemes in order to evaluate the performance of our FQGwS scheme in terms of delay and packet loss (PL) for the VANET layer and RSS for the LTE-advanced layer. Other clustering and CH selection algorithms will be considered in our future works to investigate the adaptability of our FQGwS algorithm.

Simulation scenarios have been repeated several times by varying the source vehicle velocity. In each simulation scenario, after computing input parameters, the source vehicle computes the decision of the gateway based on the traffic class to be transmitted to the infrastructure using the Matlab fuzzy module we have implemented. The generated output of the FQGwS decision is afterward simulated in NS2.33 and compared to the default deterministic scheme. We also investigate the impact of the density of the VANET cluster on the performance of the system. We plot the figure using confidence interval (CI) with a certainty of 95%.

5.3.3.3.2. Impact of cluster density

In this section, we evaluate the impact of the cluster density on the performances of the C-DRIVE and Middle algorithms. Cluster density is defined by the number of vehicles belonging to the same cluster as the source vehicle. In Figure 5.3, we show the average delay of class 1 traffic while increasing the cluster density. The computed delay is the average of the delay of all the received packets. The figure shows clearly that with the FQGwS scheme lower delays are generated as compared to deterministic approaches, that is C-DRIVE w/o FQGwS and Middle algorithm w/o FQGwS. This is due to the fact that the selected gateway has better features of VANET and LTE-advanced interfaces than the CH. In fact, the FQGwS algorithm is decision efficient as it takes consideration at the same level of the LTE-advanced and VANET features. Thus, if the LTE-advanced link properties, that is BS load and source vehicle RSS, are acceptable, it is more suitable to have no intermediate gateways and relays for the class 1 traffic type. The reason for this is that, for delay-sensitive traffics such as alerts, voice and real-time gaming, the lower the delay, the better the QoS perceived by the users. Without the FQGwS algorithm, as the number of vehicles increases, the delay increases because the CH will handle and manage more vehicles. However, with FQGwS, the delay is almost invariant because the algorithm considers the load of the CH and the gateway candidates for gateway selection.

In Figure 5.4, we show the average delay of class 2 traffic while increasing the number of vehicles in the cluster. Without the FQGwS algorithm, as the number of vehicles increases, the delay increases because the CH will handle and manage more vehicles. However, with FQGwS, the delay is almost invariant because the algorithm considers the load of the CH and the gateway candidates for gateway selection. In the same way, the packet loss increases without the FQGwS scheme and remains constant with the FQGwS algorithm.

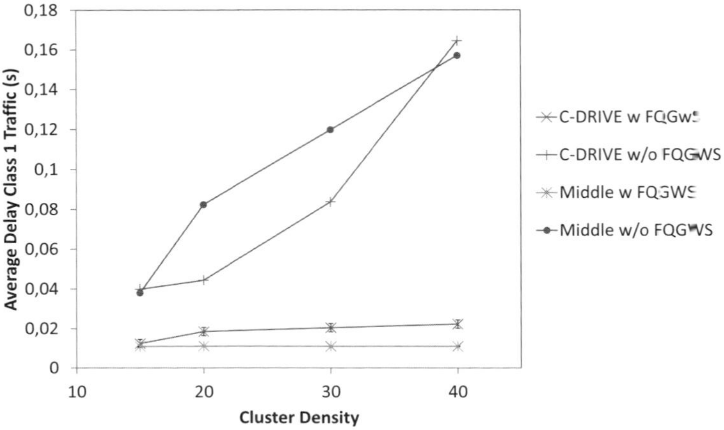

Algorithm	CI	Algorithm	CI
C-DRIVE w FQGwS	0.002104	Middle w FQGwS	0.000104
C-DRIVE w/o FQGwS	0.072776	Middle w/o FQGwS	0.023567

Figure 5.3. *Average delay versus cluster density for class 1 traffic*

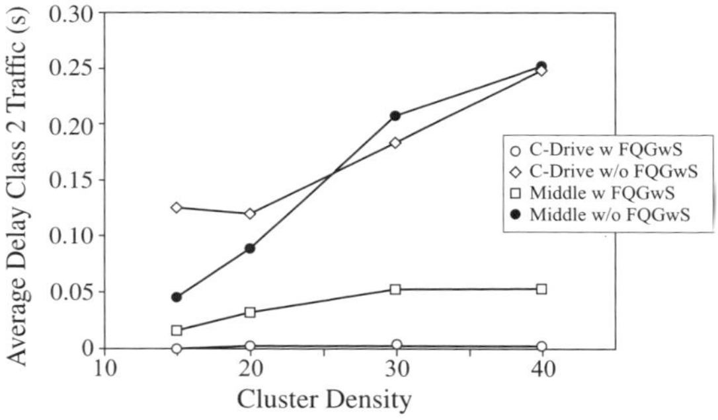

Algorithm	Cl	Algorithm	Cl
C-Drive w FQGwS	0.010996	Middle w FQGwS	0.010886
C-Drive w/o FQGwS	0.078554	Middle w/o FQGwS	0.012737

Figure 5.4. *Average delay versus cluster density for class 2 traffic*

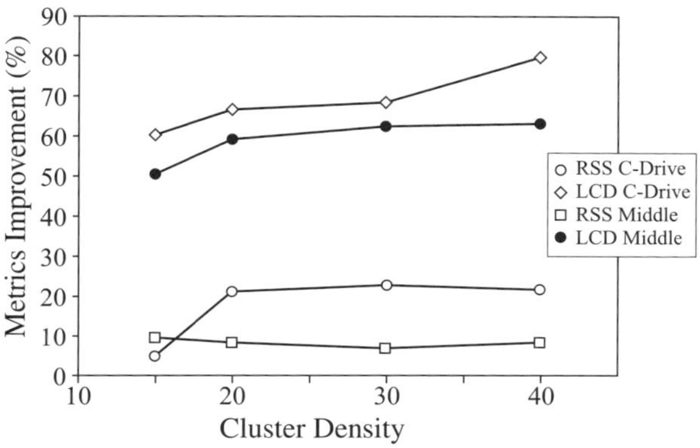

Figure 5.5. *Metrics improvement versus cluster density for class 2 traffic*

In Figure 5.5, we show the metrics improvement while comparing our algorithm to the C-DRIVE and Middle schemes alone. Two metrics are considered: RSS and LCD. The LCD metric is the link connectivity duration between the source node and the selected gateway (either the CH or the GwC). The formula we adopted to compute the metrics improvement is [5.4].

$$I_M = \frac{M_{w\ FQGwS} - M_{w/o\ FQGwS}}{M_{w/o\ FQGwS}} * 100 \qquad [5.4]$$

where M is the considered metric (RSS or LCD), I_M is the metric improvement, $M_{w\ FQGwS}$ (respectively, $M_{w/o\ FQGwS}$) is the considered metric, average value with using FQGwS algorithm (respectively, without using FQGwS algorithm).

RSS improvement (I_{RSS}) is computed using formula [5.4]. According to Figure 5.5, there is an improvement of more than 10% in terms of RSS for the Middle algorithm and about 20% for C-DRIVE. It is important to choose a gateway that has high RSS as it will interface the VANET layer with the LTE-advanced layer. An improvement of about 80% for the

Middle scheme and 90% for the C-DRIVE scheme of the LCD metric is performed with FQGwS due to the fact that our algorithm selects the gateway with a relatively low velocity and high LCD.

5.3.3.3.3. Impact of IEEE 802.11p wireless transmitting range

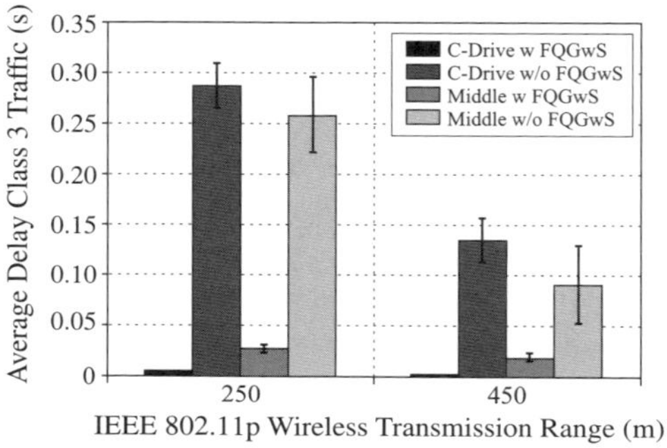

Figure 5.6. *Average delay versus IEEE 802.11p transmitting range for class 3 traffic*

Figure 5.6 clearly shows the lower delay for the FQGwS algorithm as compared to deterministic approaches that select the CH as a gateway for the source vehicle to the LTE-advanced infrastructure for class 3 traffic for both C-DRIVE and Middle algorithms. In fact, our scheme has selected a more appropriate gateway to the infrastructure that presents a lower load than the CH selected as gateway in the deterministic schemes. Therefore, with FQGwS algorithm, a lower delay is measured and fewer packets are dropped. Moreover, as the IEEE 802.11p transmitting range increases, the delay decreases. This is obvious as vehicles will become neighboring nodes and no longer be one hop from each other.

5.3.4. *Conclusions*

In this section, we proposed an original algorithm based on a fuzzy logic modeling for gateway selection from a VANET network to the LTE-advanced infrastructure. This new approach has been compared to the standard deterministic approaches that use the CH as a default gateway. Two clustering and CH selection algorithms have been studied and simulation results show that our protocol produces better results in terms of delay and packet loss than the deterministic approach for both algorithms.

5.4. Conclusion

Driven by the ever-increasing demand for high-quality services and the emergence the of new type of greedy applications (due to smart phones), a multitude of wireless technologies compete aiming to achieve user demands. The demand is such that, nowadays, efficient inter operation between those diverse technologies becomes the key requirement to overcome these challenges and accomplish users' claim. Recently, co-operative wireless communications between *ad hoc* networks and cellular systems emerge as an attractive common paradigm, especially in the ITS domain. VANET networks governed by V2V and V2I communications are crucial for ITS systems. In this chapter, we focused on studying hybrid clustered VANET network architecture that is composed of two systems: an IEEE 802.11p-based VANET network and an LTE-advanced infrastructure. We first presented the state of the art of clustering and gateway selection for traditional V2I communication in VANET networks. Then, we presented the method used to achieve a co-operative traffic transmission in a joint 4G LTE-advanced and IEEE 802.11p-based clustered VANET network. More precisely, we proposed a co-operative traffic transmission algorithm, aiming to select the appropriate gateway for a given source vehicle under the scope of V2I communications. This is a QGwS where the decision of the gateway depends on the traffic to be transmitted to the infrastructure.

5.5. Bibliography

[ALM 10] ALMALAG M.S., WEIGLE M.C., "Using traffic flow for cluster formation in vehicular *ad hoc* networks", *Proceedings of the IEEE 35th Conference on Local Computer Networks (LCN)*, pp. 631–636, October 2010.

[AMD 09] AMDOUNI I., FILAII F., "Intelligent strategies of access point selection for vehicle to infrastructure opportunistic communications", *Proceedings of the Vehicular Networking Conference (VNC), IEEE*, pp. 1–8, 2009.

[AND 10] ANDRE C., SUSANA S., PETER S., "On the connection availability between relay nodes in a VANET", *Proceeding of the IEEE Globecom*, pp. 181–185, 2010.

[BAS 99] BASAGNI S., "Distributed clustering for *ad hoc* networks", *Proceedings of the 4th International symposium on Parallel Architectures, Algorithms and Networks, I-SPAN*, pp. 310–315, 1999.

[BAS 01] BASU P., KHAN N., LITTLE T., "A mobility based metric for clustering in mobile *ad hoc* networks", *Proceedings of the Conference on Distributed Computing Systems*, pp. 413–418, 2001.

[BCH 10] BCHINI T., TABBANE N., TABBANE S., *et al.*, "Fuzzy logic based layers 2 and 3 handovers in IEEE 802.16e network", *Journal of Computer Communications*, vol. 33, no. 18, pp. 2224–2245, 2010.

[BEL 10] BELANOVIC P., VALERIO D., PAIER A., *et al.*, "On wireless links for vehicle-to-infrastructure communications", *IEEE Transactions on Vehicular Technology*, vol. 59, no. 1, pp. 269–282, 2010.

[BEN 11] BENSLIMANE A., TALEB T., SIVARAJ R., "Dynamic clustering-based adaptive mobile gateway management in integrated VANET-3G heterogeneous wireless networks", *Proceedings of the IEEE JSAC*, 2011.

[DRO 11] DROR E., AVIN C., LOTKER Z., "Fast randomized algorithm for hierarchical clustering in vehicular *ad hoc* networks", *Proceedings of the Ad Hoc Networking Workshop (Med-Hoc-Net) 2011, The 10th IFIP Annual Mediterranean*, 2011.

[ELG 08] ELGHNAMI D., KWON T., HAFID A., "Grids: geographically repulsive insomnious distributed sensors–an efficient node selection mechanism using passive clustering", *Proceedings of the IEEE WIMOB*, pp. 241–246, 2008.

[FAN 08] FAN P., SISTLA P., NELSON P., "Theoretical analysis of directional stability-based clustering algorithm for VANET", *Proceedings of the ACMVANET*, pp. 80–81, 2008.

[FU 08] FU Y., CHAN K.-M., TAN K.-S., *et al.*, "Multi-metric gateway discovery for MANET", *IEEE VTC Spring*, pp. 777–781, 2006.

[GPP 11] 3GPP, "Ts 36.423 v10.1.0", March 2011.

[JAN 95] JANG J.-S., SUN C.-T., "Neuro-fuzzy modeling and control", *Proceedings of the IEEE*, vol. 83, pp. 378–406, 1995.

[JER 07] JERBI M., SENOUCI S., RASHEED T., *et al.*, "An infrastructure- free traffic information system for vehicular networks", *Proceedings of the 66th IEEE VTC*, pp. 2086–2090, 2007.

[KWO 03] KWON T., GERLA M., VARMA V., *et al.*, "Efficient flooding with passive clustering–an overhead-free selective forward mechanism for *ad hoc*/sensor networks", *Proceedings of the IEEE*, vol. 91, no. 8, pp. 1210–1220, 2003.

[KUK 09] KUKLINSKI S., WOLNY G., "Density based clustering algorithm for VANETs", *International Journal of Internet Protocol Technology*, vol. 4, pp. 149–157, 2009.

[LIA 10] LIAO K.-H., WANG C.-H., CHU K.-C., *et al.*, "A dispersion sorting mechanism for vehicular *ad hoc* network", *Proceedings of the International Conference on Communications and Mobile Computing (CMC)*, pp. 517–521, 2010.

[LUO 10] LUO Y., ZHANG W., "A new cluster based routing protocol for VANET", *Proceedings of the Networks Security Wireless Communications and Trusted Computing (NSWCTC), Second International Conference*, pp. 176–180, 2010.

[MAN 09] MANOHARAN R., RAJARAJAN S., SASHTINATHAN S., *et al.*, "A novel multi-hop B3G architecture for adaptive gateway management in heterogeneous wireless networks", *Proceedings of the 5th IEEE WiMob*, 2009.

[MAS 09] MASLEKAR N., BOUSSEDJRA M., MOUZNA J., *et al.*, "Direction based clustering algorithm for data dissemination in vehicular networks", *Proceedings of the Vehicular Networking Conference (VNC), IEEE*, pp. 1–6, 2009.

[MIS 11] MISIC J., BADAWY G., MISIC V., "Performance characterization for IEEE 802.11p network with single channel devices", *IEEE Transactions on Vehicular Technology*, vol. 60, no. 4, pp. 1775–1787, 2011.

[RAM 02] RAMALINGAM A., SUBRAMANI S., PERUMALSAMY K., "Associativity based cluster formation and cluster management in *ad hoc* networks", *Proceedings of the 9th International Conference on High Performance Computing*, 2002.

[SAL 09] SALHI I., CHERIF M., SENOUCI S., "A new architecture for data collection in vehicular networks", *Proceeding of the IEEE International Conference on Communications, ICC*, pp. 1–6, 2009.

[SET 08] SETIAWAN F., BOUK S., SASASE I., "An optimum multiple metrics gateway selection mechanism in MANET and infrastructured networks integration", *Proceedings of the IEEE WCNC*, pp. 2229–2234, 2008.

[VAN] VanetMobiSim project, available at http://vanet.eurecom.fr/.

[WAN 08] WANG Z., LIU L., ZHOU M., *et al.*, "A position based clustering technique for *ad hoc* inter–vehicle communication", *IEEE Transactions on Systems, Man and Cybernetics–Applications and Reviews*, vol. 38, pp. 201–208, 2008.

[ZHI 12] ZHIOUA G.E.M., LABIOD H., TABBANE N., *et al.*, "An efficient qos based gateway selection algorithm for VANET to LTE advanced hybrid cellular network", *Proceedings of the ACM MSWiM*, pp. 353–356, 2012.

Chapter 6

Synthetic Mobility Traces for Vehicular Networking

6.1. Introduction

During the last decade, the research on vehicular networks has led to the proposal of countless network solutions specifically designed for vehicular environments. Such solutions encompass all network paradigms, with original algorithms and architectures proposed for next-generation roadside cellular access networks, pure vehicular *ad hoc* networks and opportunistic disruption-tolerant vehicular networks.

When such vehicular network solutions target vast deployments that comprise tens, hundreds or even thousands of cars, simulation is the means of choice for their performance evaluation. Indeed, large-scale experimental test beds are too expensive and complex to scale to more than a few communication-enabled cars, while comprehensive

---

Chapter written by Sandesh UPPOOR, Marco FIORE and Jérôme HÄRRI.

analytical approaches are often untraceable from a mathematical viewpoint.

The reliability of simulative evaluations has thus emerged as a critical aspect to the performance analysis of vehicular networks. A major role in achieving such a reliability is played by the representation of car mobility in simulation. As a matter of fact, the road traffic dynamics are the most distinctive feature that tells vehicular environments apart from other mobile scenarios. The vehicular mobility determines the node (from an *ad hoc* perspective) or user (from the point of view of the cellular access infrastructure) movement, which, in turn, drives the topology of the network (*ad hoc* scenario) or the user demand (cellular scenario).

As repeatedly proven in the past [BAI 03, FIO 08, VIR 11], the movement of vehicles can dramatically affect the behavior of network protocols, and an incorrect representation of car traffic can lead to misleading performance results and wrong conclusions, even if the network-level simulation is flawless. As a result, it is today widely acknowledged that, for the results of a vehicular network simulative campaign to be credible, the latter must build on mobility traces that faithfully capture the unique macroscopic and microscopic dynamics of car movement patterns.

The safest approach to the generation of mobility traces for vehicular network simulation would then be to extract mobility patterns from a direct monitoring of real-world road traffic. Indeed, projects aiming at tracking and logging car traffic in the real world first appeared during the first half of the past decade [JET 03, BUR 06] and have since then grown in number and scale [HUA 07, SED 08, DOE 10, ZHA 10, YUA 11]. However, the resulting traces are today limited to subsets of specific vehicles, for example, buses or taxis, and typically exhibit reduced temporal detail, that is order-of-minute position information updates. As a result, these traces are mainly employed for the performance evaluation of applications based on delay-tolerant or opportunistic networking paradigms. However, it is clear that significantly higher penetration rates and time granularity are needed for more general use cases. An interesting breakthrough in the collection of real-world vehicular movement data could be provided by the diffusion of recent navigation systems [TOM 10, MEI 11], which periodically communicate car position information to traffic management centers. Although this may greatly help real-world traffic tracking to scale up, it is easy to foresee that market rules and privacy concerns will hinder the public disclosure of such data, similarly to what happens today with the mobile access network logs of telecom operators.

As a result, synthetic vehicular mobility traces have emerged as the most viable solution to represent node or user dynamics in network simulation. The generation process of such synthetic traces has drastically improved over the last few years. The simplistic stochastic models employed in early works [DAV 00, BAI 03] have been replaced by random mobility over realistic road topologies [SAH 04], at first, and by microscopic vehicular models borrowed from transportation research [JAA 05] later on. These features were then included in dedicated simulation environments and integrated with road signalization [KRA 02, HÄR 11]. Since then, vehicular mobility simulators have been increasing their complexity and features, allowing us to accurately simulate the individual movement of vehicles over realistic road topologies [HÄR 09].

In this chapter, we will provide a comprehensive survey on the generation of vehicular mobility traces for network simulation. An introduction to the process of generating synthetic mobility traces is presented in section 6.2, where we also briefly outline the history of synthetic vehicular mobility generation. Section 6.3 then provides a taxonomy of the most important tools employed in the process above, that is the microscopic-level simulators that are available today for the generation of vehicular mobility traces. The outcome of the process, that is the mobility traces that are publicly available for the simulation of vehicular networks, are then presented in section 6.4.

6.2. Generation process

When simulating intervehicular applications, the crucial role of synthetic vehicular mobility traces has been illustrated by multiple studies [FIO 08, SOM 11]. As illustrated in Figure 6.1, the generation of realistic synthetic traces of vehicular mobility builds on three major components mutually influencing each others, that is the road topology information composing *motion constraints*, and the microscopic flow description and the macroscopic traffic description, composing *traffic demands*. Traffic demands provide a description of the desired mobility, whereas motion constraints limit them. In this section, we introduce such components, outlining their importance to the realism of the resulting movement trace. Discussing the different components in the order above also gives us the opportunity to review the evolution that the simulation of road traffic for vehicular networking has undergone during the last decade.

6.2.1. *Road topology database*

Early works on vehicular networks used to employ very intuitive representations of car mobility. The very first simulations of vehicular networks constrained the car movement over simplistic graphs, so as to mimic traffic dynamics over a street layout. These graphs were typically regular grids [DAV 00, BAI 03], also referred to as Manhattan scenarios and depicted in Figure 6.2(a) or simple user-defined layouts [TIA 02], as shown in Figure 6.2(b). Automatic road topology generation was also explored, by leveraging clustered Voronoi tessellation [HÄR 11], which led to layouts as shown in Figure 6.2(c). However, all these approaches did not yield a sufficient level of realism and were soon abandoned.

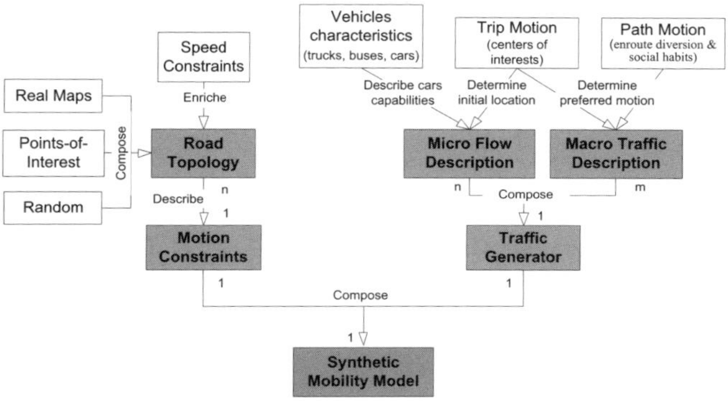

Figure 6.1. *Concept map for synthetic vehicular traces, where the two main blocks are enriched with added blocks and features providing an increased accuracy*

Real-world maps then started to be used to model the movement-constraining graph. The U.S. Census Bureau TIGER database [BUR 12], providing complete topological information on the whole US road network and being freely accessible, became especially popular to that end. Notable works that used the TIGER database include [SAH 04, NAN 06], while an example of TIGER road layout is shown in Figure 6.2(d). Another source of road topology information that became quite popular was Ertico's geographic data file (GDF) database [ERT 12]. The GDF format was more complete than the TIGER format, but the two databases were complementary, as the Ertico database provided topology information of European cities. The GDF format was employed, for example, in [HÄR 11]. A sample road topology is shown in Figure 6.2(e).

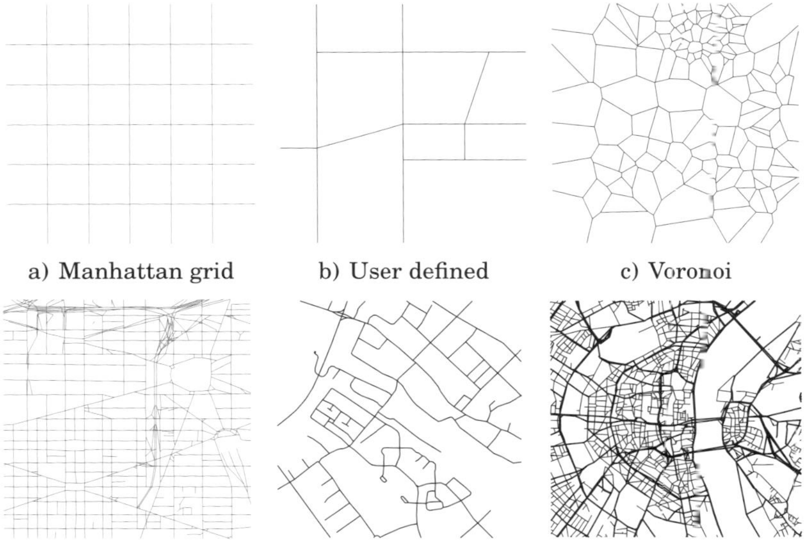

a) Manhattan grid b) User defined c) Voronoi

d) TIGER e) GDF f) OSM

Figure 6.2. *Road topology database examples*

More recently, the success of the OpenStreetMap (OSM) [OPE 12] initiative has granted to the OSM format a *de facto* standard status. The OSM initiative applies the open-source concept to road cartography, and thus provides freely exportable maps of cities worldwide, which are contributed and updated by a vast user community. Map information includes roads, railways, buildings and points of interests (PoI) such as parks, commercial centers, leisure centers and commercial activities. The OSM road information is generated and validated by the user community through satellite imagery and GPS traces, and is commonly regarded as the highest-quality road data publicly available today. Indeed, the accuracy of OSM street layouts, comprising highways, major urban arteries and minor roads, often matches that of proprietary ones such as Google Maps or Mappy, especially for large cities. An example of OSM topology is provided in Figure 6.2(f).

6.2.2. *Microscopic traffic flow description*

The second key component in the generation of vehicular movement traces is the microscopic-level mobility model. This model controls the acceleration (possibly deceleration) of an individual car at each time instant, thus determining its speed over time. Diverse microscopic mobility models can have very different levels of detail. In the following, we provide an overview of the evolution of microscopic mobility modeling for vehicular network simulation over the last few years. For a more complete discussion, we refer the reader to [FIO 09].

Early works on the simulation of vehicular networks employed *stochastic* microscopic mobility models that, in their simplest implementation, assigned to each vehicle a random speed in a predetermined speed range. The vehicle then traveled at a constant speed during its whole trip. Basic stochastic models were employed, for example, in [DAV 00, TIA 02, SAH 04]. In the latter work, the speed range was made road dependent, and obtained as $[v_{\max} - \delta, v_{\max} + \delta]$, where $v_{\max}$ was the maximum speed allowed on the current road, and δ was a tunable value.

These models are clearly not representative of a real-world driver behavior, since they completely disregard the fact that the speed of each vehicle depends on the presence of other vehicles nearby at least. As a result, vehicles can, for example, travel over one another and cross an intersection at the same time. A more complex stochastic model was introduced in [BAI 03], so as to avoid the possibility of vehicles traveling over each other. The proposed model, called the *Freeway* model, still employed random speed decision, but forced each driver to maintain a speed lower than that of the preceding vehicle. However, the Freeway model was proven to yield properties very far from the minimum requirements demanded for a realistic microscopic mobility simulation [FIO 08].

A few microscopic mobility models were proposed by leveraging *traffic stream* properties of road traffic. These models looked at vehicular mobility as a hydrodynamic phenomenon and related the three fundamental variables of velocity v (in km/h), density ρ (vehicles/km) and flow q (in vehicles/h), according to the fundamental equation:

$$\frac{\partial \rho}{\partial t} = -\frac{\partial q}{\partial x} = -\frac{\partial (\rho v)}{\partial x}. \quad [6.1]$$

As a result, the speed of a vehicle was determined according to the road traffic intensity over the road segment it was travelling on. Examples of traffic stream models employed in the vehicular networking literature can be found in [SES 92, RUD 02]. Although they represent an improvement with respect to stochastic models, traffic stream models yield a limited level of detail, for example forcing all the vehicles on the same road segment to travel at the exact same speed. As such, traffic stream models were also proven not to be a solution for the simulation of vehicular networks [FIO 08].

A significantly higher level of realism is provided by *car-following* models that describe the behavior of a driver depending on the state of its neighboring vehicles. Most car-following models determine the acceleration of a vehicle in relation to the state of the car in front. A general expression for a car-following model is:

$$\frac{dv_i(t)}{dt} = f(v_i(t), v_j(t), x_j(t) - x_i(t)), \quad [6.2]$$

where $v_i(t)$ and $x_i(t)$ are the speed and position of vehicle i at time t. The equation allows us to compute the acceleration of i, depending on the absolute and relative speed of both vehicles, as well as their distance $x_j(t) - x_i(t)$. Examples of car-following models employed in the simulations of vehicular networks abound, and include the intelligent driver model (IDM) [TRE 00] employed, for example, in [JAA 05, HÄR 11, BAU 08], the Krauss model [KRA 97] adopted, for example, by [BRE 01, KAR 07] or the Gipps model [GIP 81] used in [FRA 06]. All these models have been validated through standard tests proposed by the transportation research community.

An alternative to car-following modeling, that is equivalently valid, is represented by *cellular automata* models. These models discretize time (by assuming a fixed-length time step), space (by fragmenting roads into cells each hosting a single car at most) and speed (by considering a finite state of instantaneous speeds). The movement of cars is then described, at each time step, as a shift of finite states (i.e. vehicles with an associate speed) along a one-dimensional lattice of cells. Different cellular automata models implement different rules to realize the shift: a widely adopted model is the Nagel-Schreckenberg model [NAG 92], which is used for the simulation of vehicular networks in [FRA 06].

All the models discussed to this point deal with *intraflow* microscopic mobility, that is they can be used to determine the movement of vehicles over a single unidirectional lane. However, in most real-world scenarios, *interflow* interactions must be modeled as well. The latter occur in highway traffic, when, for example, one vehicle overtakes the other in front (and thus changes lane, merging into a new flow) or joins the traffic from a ramp. Similarly, when simulating urban traffic, flows constantly merge at intersections or roundabouts. Thus, modeling intraflow mobility is critical to achieve a complete road traffic representation in vehicular network simulation.

Models have thus been proposed that complement the intraflow descriptions presented above, by adding rules to manage the movement of a vehicles from one flow to another. These interflow models are typically built over realistic car-following or cellular automata models, and extend them so as to mimic lane changing and ramp in-flow [TRE 02, KRA 98, NAG 98], as well as intersection management in the presence of road signalization (e g. traffic lights, stop or yield signs) [HÄR 11]. A detailed presentation of all these interflow models is out of the scope of this chapter, and we once again point the interested reader to [FIO 09] and references therein for a thorough discussion.

6.2.3. *Macroscopic road traffic description*

A detailed representation of the road topology and a faithful driver behavior model are not sufficient to the generation of a realistic synthetic trace of road traffic. Indeed, the first two components discussed above allow vehicles to move on a real-world street layout with an accurate microscopic behavior, that is stopping at red traffic lights, respecting speed limits, performing overtakings and avoiding accidents. However, an important question remains unanswered: *where should vehicles go?* In other words, the missing component is the macroscopic description of road traffic that determines when vehicles start their trips and from which location, which route they choose, and where they finally stop.

The macroscopic description is critical to the realism of the resulting trace, since it is easy to understand that it determines the volume of traffic in the scenario, as well as its evolution over time. In addition, the start and stop locations of trips and the routes followed by each vehicle in the real world guide the planning of the road topology, which, as a consequence, is designed to accommodate the real-life traffic flows. Therefore, injecting unrealistic (e.g. random) flows in the road topology risks leading to large traffic jams in some areas and road underutilization in others.

In fact, the availability of accurate road topology information, discussed in section 6.2.1, and of detailed intra- and interflow mobility models, as from section 6.2.2, makes the description of macroscopic traffic the real challenge on the path toward realistic synthetic traces of vehicular mobility. To faithfully mimic macroscopic traffic flows along highway networks and across metropolitan areas, we must correctly identify the *traffic demand*, that is the start time, the origin and the destination of each car trip in the simulated region, represented as a so-called Origin/Destination (O/D) matrix. Then, an appropriate *traffic assignment* algorithm needs to be run on the O/D matrix, so as to identify the realistic route followed by each driver to reach his/her destination.

Most of the simulative evaluations in the vehicular networking literature neglect the importance of the macroscopic traffic description, and assume random traffic demands, that is O/D matrices that contain random locations and times, or fixed probabilistic turns at each intersection. Also, the shortest path is the common solution to the traffic assignment problem. Such an approach easily leads to unrealistic flows and thus strongly biased simulation results.

However, as the awareness on the importance of macroscopic modeling grows, solutions start to be proposed and adopted in order to generate traces that are also realistic from the viewpoint of large-scale traffic flows. In section 6.4, we will thus classify the datasets currently available for the simulation of vehicular networks according to the macroscopic description they employ.

6.3. Mobility simulators

We describe in this section different tools available to the community to extract synthetic mobility traces for vehicular networking. A larger coverage on vehicular mobility simulators may be found in [HÄR 10, HÄR 09].

6.3.1. *Microscopic traffic simulators*

Mostly developed for urban traffic engineering, microscopic traffic simulators such as PARAMICS [PAR 12], CORSIM [TSI 12], VISSIM [VIS 12a], TRANSIMS [TRA 12], AIMSUN [AIM 12], VanetMobiSim [VAN 12] or SUMO [SUM 12] are able to model individual urban traffic at a microscopic level. Yet, not all of these simulators can be used straightaway for vehicular networking, as either synthetic traces cannot be exported or open Application Programming Interface (API) are not available to interconnect them with a network simulator. The semicommercial simulators VISSIM and AIMSUN provide support for trace extractions and interconnection. In the interest of comparability of research results, it is however more beneficial to use readily available free and open source software simulators, such as SUMO. SUMO is capable of extracting synthetic traces in the shape of data files and provides external control through an open API called TraCI.

From a macroscopic road traffic level (topological maps and traffic models), they mostly all support similar features, but it is at the microscopic level that they differ. For instance, TRANSIMS and AIMSUN are based on an extension of the Gipps microscopic flow equations [GIP 81], whereas VISSIM is based on a psychophysical flow model from Wiedemann [WIE 74]. SUMO initially supported only the Krau flow model [KRA 97] and its extensions, but has been recently been extended to include the same complex Gibbs model called IDM from Treiber *et al.* [TRE 00] as VanetMobiSim.

Microscopic traffic simulators represent the most used simulators to simulate vehicular networks, as they are capable of modeling individual mobility at a high level of precision. It is also this feature that made microscopic traffic simulators popular for the design and evaluation of traffic safety applications based on vehicular communications. The scale is their limitation, as modeling individual and strongly correlated mobility at a microscopic level requires a significant computational capability. Nevertheless, simulators such as SUMO or VISSIM are capable of simulating up to 10,000 vehicles faster than or equal to real-time, and SUMO has been the preferred choice for the open-source intelligent transportation system (ITS) simulation platform iTETRIS [ITE 11].

6.3.2. *Mesoscopic traffic simulators*

One typical approach to simulate traffic at a mesoscopic level is to model road elements as a queue and intersections as nodes. The vehicular dynamics are represented by queue parameters, such as the queue capacity (the maximum bumper-to-bumper vehicles allowed on a road element) and the maximum flow of the queue. The node parameters represent the intersection scheduling policy, such as the intersection type, level of priority and access control to connecting queues. The simulators Matsim [MAT 12] and FastTrans [FAS 11] are two popular tools in this category. Both simulators support the import of detailed topological maps and detailed added features, and also include efficient traffic models (macroscopic) such as advanced Dijikstra routing and origin-destination matrices. Both are capable of simulating and extracting synthetic traces of individual cars up to a city-wide scale. FastTrans also provides parallelization on multiprocessors for a higher scalability.

The mesoscopic traffic simulators represent the best trade-off between granularity and scalability, as they can simulate very large scenarios, but still provide the detailed mobility values of each vehicle. For instance, mesoscopic simulators are a preferred choice to simulate large-scale impacts of vehicular communication to improve traffic efficiency. Mesoscopic simulators are yet not recommended for the evaluation of vehicular networking for traffic safety applications, as they do not provide a sufficient granularity, in particular, between vehicles located in the same mesoscopic queue.

6.3.3. *Macroscopic traffic simulators*

Traffic simulated at a macroscopic flow level does not consider the mobility parameters of a specific car but instead quantities of macroscopic meanings such as flow, speed or density. Inspired by fluid theory, macroscopic flows have the advantage of a reduced computational complexity compared to the microscopic flows but they can still realistically model macroscopic quantities. The most well-known simulator capable of simulating vehicular traffic at a macroscopic scale is VISUM from PTV [VIS 12b]. As other simulators, it is capable of importing precise topological maps and can configure the traffic light policies at a very precise level. It is typically used by urban planners and traffic departments at the city and regional level to evaluate traffic congestions.

The synthetic traces obtained by macroscopic simulators cannot be used for vehicular networking directly, as they do not provide detailed mobility values of single entities, but instead aggregated measures such as density and speed. Macroscopic simulators are also unable to model mobility changes or personalized navigation of single vehicles that is typically found in ITS applications for traffic efficiency. Yet, aggregated mobility values are crucial for the design of efficient vehicular networking solutions, for instance, considering that aggregated vehicular density in various streets are indicators of vehicular connectivity. Also, macroscopic synthetic traces are often used to calibrate the microscopic traffic simulators, such as SUMO. Within the iTETRIS project [ITE 11], for instance, traffic in the city of Bologna has been simulated using SUMO, but during the calibration step, synthetic traffic flows extracted from VISUM have been used.

6.3.4. *Interactions between simulators*

Obtaining close-to-reality synthetic traffic traces and using it for vehicular networking requires using multiple simulators. Traffic calibration first requires multiple sources of data that come from external sources. Second, vehicular networking needs to be able to use the mobility information from the simulators. For the sake of clarity in this chapter, we will differentiate between the use of multiple traffic simulators to calibrate one or multiple simulators to obtain synthetic traces, and the use of multiple types of simulators for vehicular networking. However, conceptually speaking, both categories fall into the same domain relating the interactions between simulators.

6.3.4.1. *Calibrating simulators*

The first step to design and obtain synthetic traffic traces is to select the traffic simulator fitting best. Yet, as any tool, a simulator needs to be fed with parameters, such as topological map, traffic demand, traffic lights policy or driver models. Such data may either be obtained from statistical behaviors or extracted from dedicated simulators.

Considering the case of the calibration of the city of Bologna, as conducted during the iTETRIS project [ITE 11], the selected traffic simulator was SUMO. Yet, multiple various other simulators have been used to be able to calibrate the synthetic traces. Conceptually speaking, calibrating a simulator may be represented as in Figure 6.3. First, the macroscopic simulator VISUM has been configured with the topological map of the city of Bologna and from vehicular flow interactions measured at intersections. The output of VISUM took the shape of traffic volumes at each road segment that have later been imported by SUMO to calibrate its traffic demand. Second, a traffic light control simulator has been used to be able to provide a close-to-reality traffic light policy in the city of Bologna. This policy has then been integrated into SUMO. The output has been a calibrated synthetic trace of the typical traffic in the city of Bologna for 1 hour.

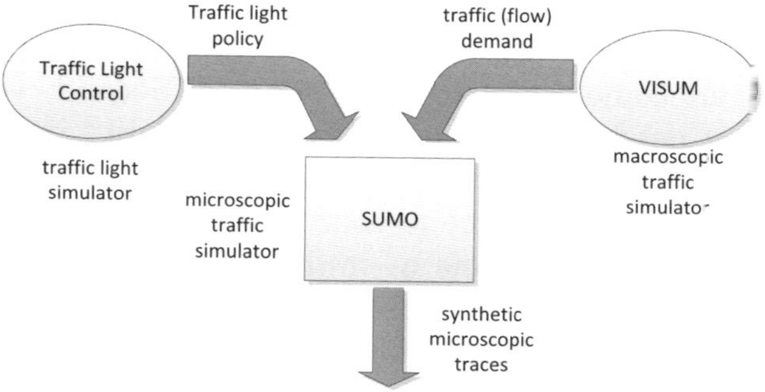

Figure 6.3. *The microscopic simulator SUMO calibrated by external simulators*

6.3.4.2. *Interaction between simulators for vehicular networking*

We previously described the different methods required to obtain close-to-reality synthetic vehicular traces. Yet, in order to be used by the vehicular networking community, these traces need to be made available to network simulators. The description of network simulators available to vehicular research is out of the scope of this chapter, but we may still cite ns-3 [NS3 12] and OMNET++ [OMN 12] as two popular simulators that are often used in research, as cited in [JOE 12].

The interaction between traffic and network simulators may be categorized into two groups as illustrated in Figure 6.4 *isolated* and *federated.* In the former case, synthetic traces are extracted from traffic simulators in the shape of mobility files that may later be integrated into the network simulator using a dedicated parser. In the latter case, a bidirectional interaction is created between two (or more) simulators to be able to exchange mobility data or to influence the control of one or the other. This approach is more complex, as an interface has to be developed to synchronize and exchange data.

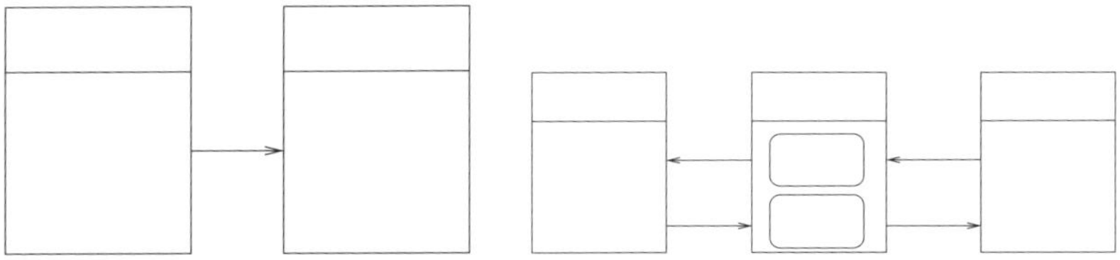

a) Isolated case b) Federated case

Figure 6.4. *Interactions between network and traffic simulators*

The *isolated* case was initially the preferred choice for network design based on small-scale random mobility, but nowadays, network simulators are able to integrate large-scale realistic mobility data files from various sources. This approach also remains a favorite choice when mobility does not need to be influenced by vehicular communication or networking. Considering applications for ITSs, mobility shall be influenced, either to avoid an accident, or to reduce congestion. So, the *federated* case is mostly used in that case. Several integrated tools have been created and made available to the community. The iTETRIS platform interlinks the simulator SUMO and ns-3 [NS3 12], the VEINS platform [VEI 12] interlinks SUMO and OMNET++ [OMN 12], and VSIMRTI [VSI 12] provides APIs to interlink jist/swans [SWA 12] and VISSIM. iTETIS and VEINS are both open source, but VsimRTI only provides open APIs without releasing the code of its core.

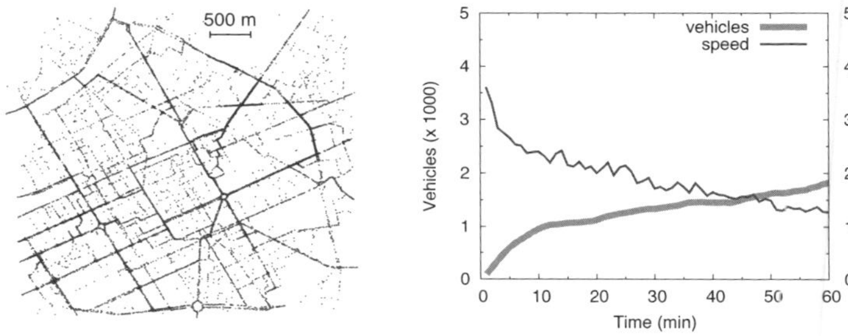

a) Traffic snapshot b) Traffic volume and speed

Figure 6.5. *Turin scenario: city center, sparse traffic*

6.4. Mobility traces

In this section, we will overview the body of work on vehicular mobility traces for network simulation. We categorize the datasets based on the nature of their macroscopic traffic data, that is the sources employed to determine the time and routes of trips traveled by individual vehicles in the dataset. The most relevant features of the different mobility traces we discuss in the following are summarized in Table 6.1.

Type/Name	Simulator	Duration (h)	Area (km^2)	Vehicles	Granularity (s)	Roads
Perception						
Turin [BAR 09]	SUMO	1	20	1.500	1	Major/minor
Porto [CON 08]	DIVERT	0.3	62	5.000	1	Major/minor
Zurich [BAU 08]	GMSF	0.5	9	420	1	Major/minor
Measurements						
Karlsruhe [FIO 12]	VanetMobiSim	0.7	21	2.000	1	Major/minor
Berlin [LOC 03]	Videlio	14	21.56	955	1	Major/minor
Photography						
Porto [FER 09]	-	-	41.3	10.566	snapshot	All
Detectors						
Berkeley [BAI 09]	-	72	-	-	60	Single freeway
Toronto [BAI 09]	-	24	-	-	20	Single freeway
Bologna [ITE 11]	SUMO	1	20.6	10.333	1	Major/minor
Luxembourg [PIG 11]	SUMO	12	1.700	150.000	1	Major
Survey						
Portland [WAG 99]	TRANSIMS	0.25	21	16.529	1	Major/minor
Canton of Zurich [RAN 03]	MMTS	6+6	65.000	260.000	1	Major
Cologne [UPP 11]	SUMO	24	400	770.000	1	All

Table 6.1. *Major features of publicly available vehicular mobility traces*

6.4.1. *Perception*

A simple way to fill the O/D matrix and determine the traffic assignment is to employ one's perception of the road traffic. The authors [BAR 09] leveraged their knowledge of the traffic in Turin, Italy, so as to determine the macroscopic demand injected in different areas of the city. The traces were generated using OSM as the road topology database and SUMO as the microscopic traffic simulator. Figures 6.5 and 6.6 refer to the Turin city center and Turin outskirts scenarios, respectively. In both figures, the left plots portray a snapshot of the vehicular mobility, so as to provide an intuitive glimpse of the mobility reproduced by each trace. Also, dot brightness is used to distinguish fluid traffic conditions from traffic congestion: light dots denote high average speeds, whereas dark ones indicate speeds close to zero. We can observe heavy localized traffic jams in Figure 6.5(a): given the sparse traffic conditions assumed in this scenario, congestion is an artifact of the incorrect traffic demand modeling rather than a representation of a real-world phenomenon. The snapshot in Figure 6.6(a) depicts instead a much simpler road topology, where traffic is fluid along the few major roads and absent elsewhere. Congestion is only recorded around the main roundabout in the northern region of the scenario. Figure 6.5(b) and Figure 6.6(b) show instead the time evolution of the traffic volume and average speed, for the same two scenarios. We can observe that, notwithstanding the limited duration of the traces, none of them is in a stationary condition. In both cases, the number of vehicles in the region tends to grow, while speeds decrease. The significant difference between the traffic conditions at the beginning and at the end of the traces is another evidence of the difficulty of calibrating the traffic demand based on pure perception. As an example, in Figure 6.5(b), the average speed starts at 30 km/h (that, considering the presence of traffic lights, denotes a completely empty road network) and progressively drops to 10 km/h (that is instead a symptom of major congestion) in less than 1 hour. Such a phenomenon would be hardly observed in the real world, in the presence of the sparse traffic assumed in the trace.

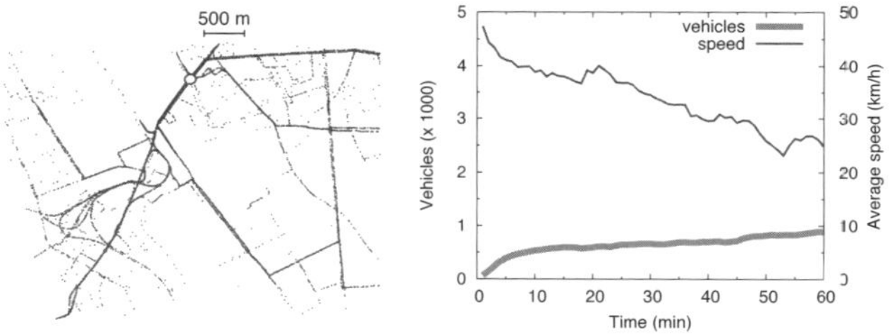

a) Traffic snapshot b) Traffic volume and speed

Figure 6.6. *Turin scenario: city outskirts, dense traffic*

We remark that a similar macroscopic approach is also adopted in [CON 08], although a different microscopic simulator framework, named DIVERT, is employed. The city concerned is in this case Porto, in Portugal.

The technique used to generate mobility traces of the region of Zurich, Switzerland, in [BAU 08], is instead slightly different. Apart from the microscopic simulation environment adopted, that is this time Generic Mobility Simulation Framework (GMSF), and the road topology information, retrieved from the Swiss Geographic Information System (GIS), the macroscopic traffic flows are injected assuming that larger roads attract more traffic. The features of the resulting mobility are reported in Figure 6.7, for one of the scenarios in [BAU 08], namely that representing the city center (the others refer instead to suburban and rural areas). From the traffic snapshot, in Figure 6.7(a), we can observe that, in the small region compassed by the simulation (less than 10 km^2), the traffic tends to be consistently slow over the whole street layout. Such a uniformity is also present in time, as the number of vehicles in the region and the average speed are constant over the 20 min of duration of the trace. Overall, the Zurich city center trace exhibits speeds exceedingly low with respect to the traffic volume considered. More importantly, the mobility description appears significantly limited in terms of size and temporal duration.

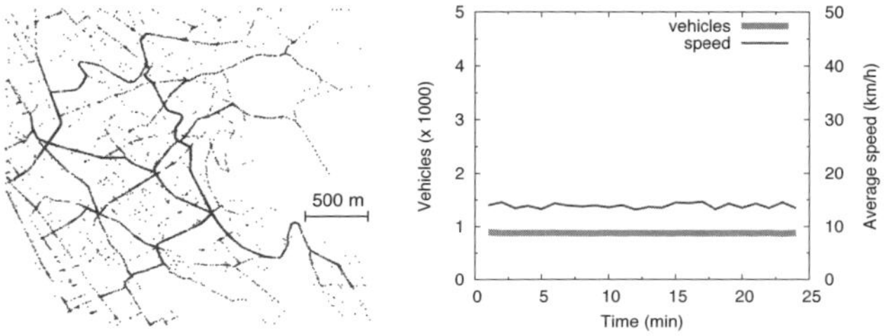

a) Traffic snapshot b) Traffic volume and speed

Figure 6.7. *Zurich scenario: city center*

In fact, a common weakness of traces built on simplistic macroscopic traffic descriptions is that they show limited extension, in both space and time. As a matter of fact, they only cover a few tens of square kilometers, and have durations ranging between 20 min and 1 hour. This limitation is a by-product of the macroscopic description: the perception approach can be applied on small areas for short time periods. However, the lack of real-world data behind the traffic demand and assignment leads to significant problems when large regions or long time periods are taken into account. Manually assigning large-scale flows so that the road topology can accommodate them is a prohibitive task. Similarly, the introduction of realistic variations of road traffic over time is hard to accomplish, if no realistic data are available to support it.

6.4.2. *Small-scale measurements*

Several works have exploited small-scale measurements, typically conducted by the authors themselves, so as to provide some simple validation to the authors' perception or to complement the latter. This is the case of the mobility trace presented in [LOC 03] that reproduces the vehicular traffic in the center of Berlin, Germany. The trace, generated with a dedicated microscopic simulator named Videlio, covers a small area of around 20 km^2 for a quite long period of 14 hours. However, the small number of vehicles (less than 1,000) present in the trace over such a long time interval raises questions on the realism of the macroscopic traffic demand employed.

Small-scale measurements are also at the base of the macroscopic modeling in the trace of downtown Karlsruhe, Germany, employed in [FIO 12]. The dataset, generated by feeding a OSM map to the VanetMobiSim microscopic mobility simulator, mimics the movement of 2,000 vehicles for approximately 40 min. The plots in Figure 6.8 show that, similarly to what observed for the Zurich traces before, the number of vehicles is quite stable, while the vehicular mobility is characterized by a generalized low speed. In fact, the average velocity recorded in the trace shows a slow but constantly decreasing trend with a steady traffic volume, a symptom of potential problems when extending the simulation duration.

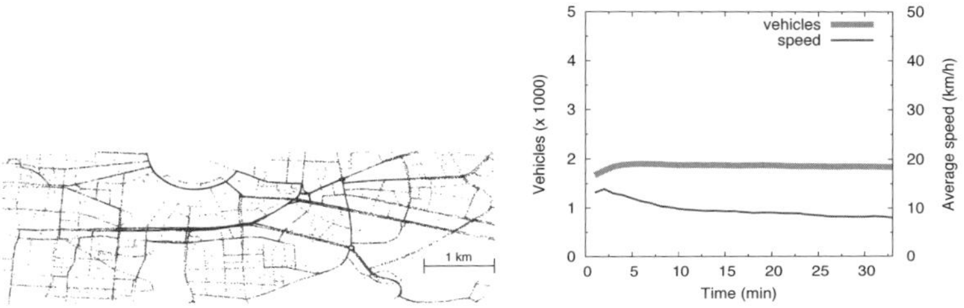

a) Traffic snapshot b) Traffic volume and speed

Figure 6.8. *Karlsruhe scenario: city downtown*

Overall, approaches based on small-scale measurement represent an appreciable effort toward the simulation of more realistic traffic demands. However, it is clear that they lack the statistical rigor needed for a realistic representation of the macroscopic traffic distribution, and, as such, can hardly capture the complexity of traffic flows in urban areas or their evolution over long time periods. For these reasons, these traces only cover modest geographical surfaces, constrained to a few tens of square kilometers, or have limited time duration, in the order of tens of minutes.

6.4.3. *Road traffic imagery*

An original solution to the derivation of the macroscopic traffic information was adopted in [FER 09], where the authors leveraged stereoscopic aerial photography in order to capture the vehicle distribution in the city of Porto, Portugal. A private aircraft was flown over the city for 2 hours in the early afternoon of a weekday, and photographs were shot from the plane every 5 s. The flight followed a parallel row pattern so as to cover the whole geographical area of 41.3 km^2 corresponding to the surface of Porto. By studying the aerial imagery, the authors were able to reconstruct a single snapshot of the positions of 10.566 vehicles in the urban area. The time error in the snapshot, due to the fact that pictures of different locations were taken at different moments, is of 23 min between two cars, on average. Although this appears as an interesting way to derive static macroscopic data, its applicability to the generation of mobility traces is not clear, due to the complexity and cost of running a large-scale aerial photography campaign for a significant period of time.

Another interesting attempt at using imagery to estimate the macroscopic behavior of car traffic has recently been presented in [THA 11]. There, the authors exploit the pervasiveness of road surveillance cameras to infer traffic densities in two large urban areas, namely London and Sydney. Although promising, this approach has not yet been used for the generation of traffic demands in mobility traces.

6.4.4. *Roadside detectors*

Induction loops, infrared counters and roadside sensors represent the traditional way to measure vehicular traffic flows in both freeways and urban road networks. The authors [BAI 09] use two sets of empirical data obtained from dual-loop and metal detectors from sections of the I-80 Freeway in Berkeley, CA, USA, and of the Gardiner Expressway in Toronto, Canada. The detector information covers a span of 24 hours and allows us to determine the per-lane intervehicle arrival time and spacing. Although the data could be fed to a microscopic simulator to derive the position of individual vehicles over time, their validity is limited to highway environments.

Roadside detectors have been employed in a urban environment within the iTetris project [ITE 11]. Vehicular mobility traces of several areas of the city of Bologna were generated accounting for macroscopic traffic data acquired through 636 induction loops spread over the road network, and complemented by user surveys on usual commuting trips. The main trace covers 20.6 km^2 in the city center for a period of 1 hour, featuring the movement of 10,333 vehicles. Because of the real-world nature of the macroscopic data they are built upon, these traces reach an unprecedented level of realism. Unfortunately, they do not cover large surfaces or long time periods.

A similar approach has been discussed in [PIG 11], where the authors calibrate the microscopic mobility simulation of the city of Luxembourg through traffic flow information gathered by the local Ministry of Transport. As such real-world data only covers major traffic arteries, it is complemented by driver routes inferred from the different nature of geographical zones in the area under study, and used to define traffic flows on medium- and small-sized roads. The resulting mobility trace covers a very large area of 1,700 km^2 and features 150,000 car trips. Although a very interesting dataset, the Luxembourg trace focuses on highway and major roads traffic, as depicted in Figure 6.9(a), and only accounts for inbound flows, that is traffic moving toward the city center. Furthermore, the dataset is limited to the morning period and yields clearly unstable traffic conditions toward the end of the simulated period, as shown in Figure 6.9(b).

6.4.5. *Sociodemographic surveys*

Sociodemographic surveys represent a significant source of information for the derivation of vehicular traffic data. Seminal work in that direction was performed in [WAG 99], where the authors generated a mobility trace whose macroscopic traffic description is obtained from knowledge of driver activity in downtown Portland, OR, USA. The resulting mobility dataset is acknowledged to be very realistic, but it only covers 15 min of vehicular movements in an small area of 21 km^2, for a total of 16.529 simulated cars. Moreover, the trace is not publicly available.

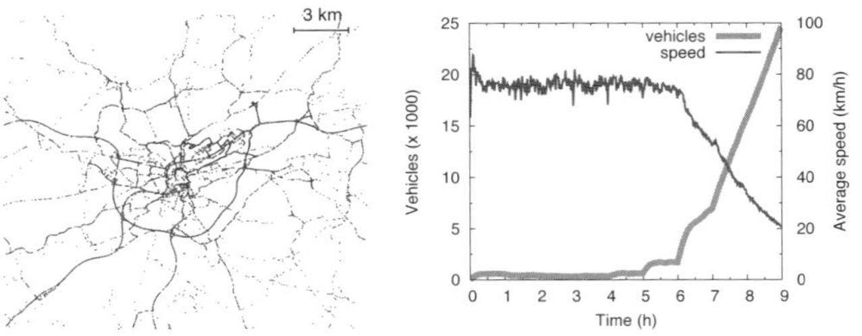

a) Traffic snapshot b) Traffic volume and speed

Figure 6.9. *Luxembourg scenario: Luxembourg city and surroundings*

The largest vehicular mobility trace generated to date is the trace reproducing the road traffic in the whole Canton of Zurich, a 65.000 km^2 region of Switzerland [RAN 03]. The mobility dataset is said to cover 24 hours, through traffic demand data obtained from the Swiss Regional Planning Authority and complemented using the 1994 Swiss National Travel Survey. The resulting trace, as well as subsets of the same, is widely employed in the vehicular networking literature. However, the size of the road topology forced the authors to limit the detail of the microscopic-level simulation. Thus, they resorted to a queue-based multi-agent microscopic traffic simulator (MMTS), significantly less accurate than standard fine-grained vehicular mobility simulators based on car-following models. Moreover, once again due to scalability reasons, the road topology was pruned down to major traffic arteries, as observable in Figure 6.10(a), that portrays a snapshot of the vehicular mobility in the most significant 400 km^2 portion of the simulated region. The selected area represents the city of Zurich and its surroundings: the low topological detail is striking, especially when comparing the figure with Figure 6.7(a), that represents that city center only, but at a much higher detail level. The light color that dominates the plot also indicates a high average speed: this is normal, since only major roads, that is highways and freeways, are represented. However, the lack of minor urban and rural roads clearly affects the completeness of the mobility description provided by the trace.

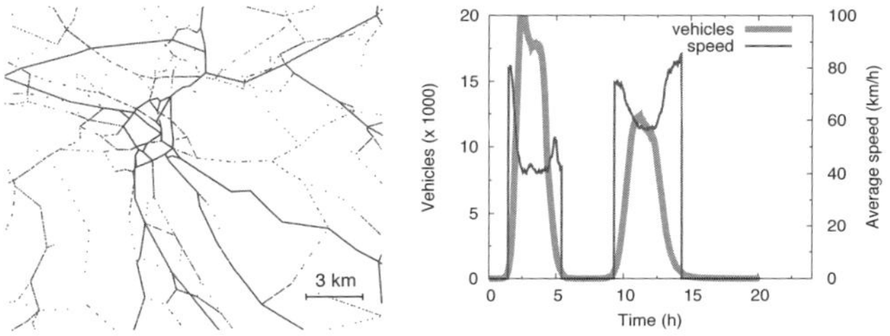

a) Traffic snapshot b) Traffic volume and speed

Figure 6.10. *Canton of Zurich scenario: Zurich city and surroundings*

Elevated average speeds are also observed in Figure 6.10(b), where, however, the more interesting aspect is the limited detail of the traffic demand. As a matter of fact, the plot evidences how the trace only considers vehicular mobility during the morning and afternoon traffic peaks, while the off-peak hours are neglected, with the traffic volume dropping to zero. The Canton of Zurich trace is therefore a clear example of the trade-off between complexity and scale in the generation of synthetic vehicular mobility traces: in order to be able to reproduce road traffic on a very large scale, we have to accept low road topology detail, low microscopic precision and approximate macroscopic description.

The vehicular mobility dataset in [UPP 11] reproduces the road traffic in the greater urban area of the city of Cologne, in

Germany. The trace was generated by extracting OSM data about 4,500 km of roads in an area of more than 400 km^2 around the urban agglomeration of Cologne. The OSM information was then fed to SUMO, and coupled with a macroscopic traffic demand provided by the TAPASCologne project, an initiative by the Institute of Transportation Systems at the German Aerospace Center (ITS-DLR). More precisely, the O/D matrix was generated though the Travel and Activity Patterns Simulation (TAPAS) methodology [VAR 06], a technique that exploits information on (1) the population, that is home locations and sociodemographic characteristics, (2) the PoI in the urban area, that is places where working and free-time activities take place, and (3) the time use patterns, that is habits of the local residents in organizing their daily schedule [HER 04]. The TAPAS methodology was applied on real-world data collected in the Cologne region by the German Federal Statistical Office, including 30,700 daily activity reports from more than 7,000 households [RIN 02, EHL 96].

The resulting TAPASCologne O/D matrix faithfully mimics the daily movements of inhabitants of the area for a period of 24 hours, for a total of 1.2 million individual trips.

The assignment of the vehicular traffic flows described by the TAPASCologne O/D matrix over the road topology was then performed by means of the Gawron algorithm [GAW 98]. This traffic assignment technique calculates the fastest route for each vehicle, and then assigns to each road segment a cost reflecting the intensity of traffic over it. By iteratively moving part of the traffic to alternate, less congested paths, and recomputing the road costs, the scheme finally achieves a so-called user equilibrium. Additionally, since the intensity of the traffic demand varies over time, the traffic assignment model must also be able to adapt to the time-varying traffic conditions. Indeed, Gawron's algorithm satisfies such a requirement, thus attaining a so-called dynamic user equilibrium. Gawron's is one the most popular traffic assignment techniques developed within the transportation research community and allows us to reach a road capacity utilization close to reality and significantly higher than that obtained with, for example, a simple weighted Dijkstra algorithm.

The resulting vehicular mobility trace is shown in Figure 6.11. Figure 6.11(a) show the large scale as well as the high level of detail of the dataset that comprises at the same time highways, rural roadways, major roads and small streets in the urban center. In addition, speeds match the expectations, since highways and peripheral roads are characterized by high velocities, while the average speed is reduced when closing down on the city center. However, no unrealistic congestion (i.e. widespread dark color) can be noticed, even if the snapshot is taken at 7.30 am, that is during the morning traffic peak. Figure 6.11(b) proves how the traffic volume is representative of the typical daily activity, with morning and afternoon traffic peaks, but a non-zero activity otherwise, including a minor peak around noon.

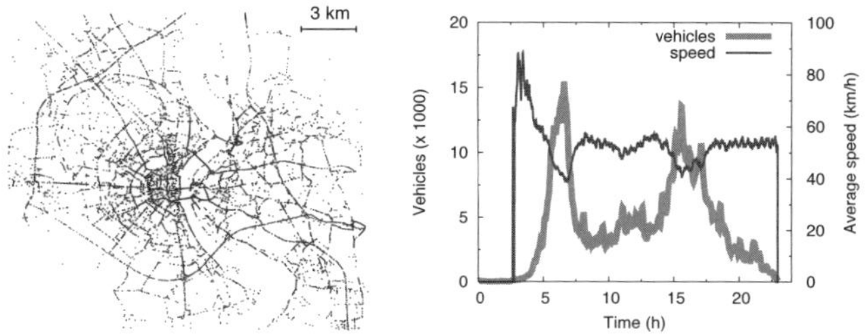

a) Traffic snapshot b) Traffic volume and speed

Figure 6.11. *Cologne scenario: Cologne city and surroundings*

6.4.6. *Discussion*

Overall, the ultimate vehicular mobility trace for network simulation should feature all of the following:

(1) representation of the completeness of the road traffic; (2) a high time granularity, the position of each vehicle being traced with at least an order-of-second precision; (3) compass very large regions (i.e. whole urban areas), with a faithful description of the road layout and signalization; (4) provide a realistic representation of the microscopic behavior of individual drivers in everyday traffic, accounting for their interactions with other drivers and their behavior with respect to road regulations; (5) be also realistic from a macroscopic point of view, by faithfully mimicking the movement of large-scale traffic flows across a metropolitan area, over long time periods.

The first two requirements rule out traces obtained though real-world tracking, as they are limited to subsets of the vehicles, that is buses or taxis, and exhibit reduced temporal detail, that is order-of-minute position information updates. These traces are today mainly employed for the performance evaluation delay-tolerant or opportunistic data exchanges, and remain interesting for applications that fit such transfer paradigms.

By then resorting to synthetic mobility traces, the third constraint can also be easily met, thanks to the availability of accurate real-world road map services. Similarly, recent vehicular mobility simulators allow us to satisfy the fourth requirement, as they implement realistic microscopic vehicular mobility models, borrowed from transportation theory and validated through comparison with real-world measurements. Today's challenge lies in fact in the last aspect, that is attaining macroscopic-level realism in the simulation of road traffic. As previously discussed, the problem is separated into two phases, that is the identification of the traffic demand and the iteration of an appropriate traffic assignment algorithm on the O/D matrix.

The synthetic traces presented in this chapter address the issue of macroscopic realism by using different data sources, as clearly outlined by the classification we proposed. Some sources are hardly credible by their own nature (e.g. perception or small-scale measurements) or not scalable due to their complexity (e.g. aerial photography). Finally, and quite unsurprisingly, the most reliable and scalable sources are those commonly employed by transportation engineering (e.g. road traffic detectors and population surveys).

However, none of the traces making use of such macroscopic traffic data sources cover large urban regions for long periods of time, and provide at the same time high microscopic-level accuracy. The Berkely and Toronto traces [BAI 09] are limited to one freeway segment. The Bologna [ITE 11] and Portland [WAG 99] datasets focus on small urban areas of a few tens of square kilometer. The Luxembourg [PIG 11] and Canton of Zurich [RAN 03] traces are closer to compassing all of the desirable features we outlined above. However, their limit lies in that they mainly model traffic on major roads, as also testified by the relatively low number of vehicles per square kilometer they comprise. The only dataset that seems to meet all the requirements is the Cologne dataset [UPP 11], which couples a large scale with a high level of detail and an accurate macroscopic traffic description.

We stress, however, that the road to the generation of ultimate synthetic vehicular mobility traces is still long. More precisely, aspects that are still to be addressed concern at the same time the scalability of microscopic simulation and the increased detail of the same. In terms of scalability, it would be desirable to obtain traces that cover whole regions, and, why not, even countries. As far as the level of detail is concerned, a highly desirable feature that is missing today is instead the timing of traffic lights in the simulation environment so that it matches the real-world configuration.

Moreover, the macroscopic traffic description remains an open problem. Apart from isolated initiatives, such as TAPASCologne, realistic road traffic demand information on vast regions is hard to retrieve. To that end, the growing trend of government bodies and local administrations making urban-related data publicly available can provide a significant breakthrough. Open access data initiatives such as Vancouver's VanMap, OpenBaltimore, London DataStore and ParisData, just to cite a few significant examples, could indeed lead to the public disclosure of road counter collections or traffic flow datasets. Such information would, in turn, grant the possibility to extract new macroscopic O/D matrices as well as validate traffic assignments.

Finally, it is important to note that random models, both at the motion constraint and traffic demand sides, have lost momentum over the past few years. Yet, they should not be ignored, as they still represent a very powerful approach for fast configuration of motion patterns, where statistical evaluations over multiple different configurations are more important than on a specific dataset. The challenge in that case is to provide random models, yet ones that are statistically representative of realistic vehicular motion patterns, such as random graphs providing statistically similar motion constraints or random traffic demands providing statistically similar macroscopic traffic patterns.

6.5. Bibliography

[AIM 12] AIMSUN, http://www.aimsun.com/wp/, 2012.

[BAI 03] BAI F., SADAGOPAN N., HELMY A., "The IMPORTANT framework for analyzing the Impact of Mobility on Performance Of RouTing protocols for Adhoc NeTworks", *Elsevier Ad Hoc Networks*, vol. 1, pp. 383–403, 2003.

[BAI 09] BAI F., KRISHNAMACHARI B., "Spatio-temporal variations of vehicle traffic in VANETs: facts and implications", *ACM VANET*, Beijing, PRC, 2009.

[BAR 09] BARBERIS C., G.M., "Large-scale simulation of V2V environments", *IEEE ICUMT*, St. Petersburg, Russia, 2009.

[BAU 08] BAUMANN R., LEGENDRE F., P.S., "Generic mobility simulation framework (GMSF)", *ACM MobilityModels*, Hong Kong, PRC, 2008.

[BRE 01] BREISEMEISTER L., Group Membership and communication in highly mobile ad hoc networks, PhD thesis, Tecnhical University of Berlin, 2001.

[BUR 06] BURGESS J., GALLAGHER B., JENSEN D., *et al.*, "MaxProp: routing for vehicle-based disruption-tolerant networks", *IEEE INFOCOM*, Barcelona, Spain, 2006.

[BUR 12] BUREAU U.C., "Topologically integrated geographic encoding and referencing (TIGER) system", http://www.census.gov/geo/www/tiger, 2012.

[CON 08] CONCEIÇÃO H., DAMAS L., FERREIRA M., *et al.*, "Large-Scale Simulation of V2V Environments", *ACM SAC*, Fortaleza, Brazil, 2008.

[DAV 00] DAVIES V., Evaluating mobility models within an ad hoc network, Master's thesis, Colorado School of Mines, 2000.

[ERT 12] ERTICO, Geographic Data Files, http://www.ertico.com/gdf-geographic-data-files, 2012.

[FAS 11] FASTTRANS, http://www.lanl.gov/programs/nisac/fasttrans.shtml, 2011.

[FER 09] FERREIRA M., CONCEIÇÃO H., FERNANDES R., *et al.*, "Stereoscopic aerial photography: an alternative to model-based urban mobility approaches", *ACM VANET*, Beijing, PRC, 2009.

[FIO 08] FIORE M., HÄRRI J., "The networking shape of vehicular mobility", *ACM MobiHoc*, Hong Kong, PRC, 2008.

[FIO 09] FIORE M., "Vehicular mobility models", *Vehicular Networks from Theory to Practice*, Chapman & Hall/CRC, 2009.

[FIO 12] FIORE M., CASETTI C., CHIASSERINI C.-F., *et al.*, "Discovery and verification of neighbor positions in mobile ad hoc networks", *IEEE Transactions on Mobile Computing*.

[FRA 06] FRACCHIA R., MEO M., ROSSI D., "VANETs: to beacon or not to beacon", *IEEE Globecom Autonet*, San Francisco, CA, USA, 2006.

[GAW 98] GAWRON C., "An iterative algorithm to determine the dynamic user equilibrium in a traffic simulation model", *International Journal of Modern Physics C*, vol. 9, no. 3, pp. 393–407, 1998.

[GIP 81] GIPPS P., "A behavioural car following model for computer simulation", *Transportation Research B*, vol. 15, pp. 105–111, 1981.

[HÄR 09] HÄRRI J., FILALI F., C.B., "Mobility models for vehicular ad hoc networks: a survey and taxonomy", *IEEE Communications Surveys and Tutorials*, vol. 11, no. 4, pp. 19–41, 2009.

[HÄR 10] HÄRRI J., "Vehicular mobility modeling for VANET", *VANET Vehicular Applications and Inter-Networking Technologies*, pp. 107-156, 2010.

[HÄR 11] HÄRRI J., FIORE M., FILALI F., *et al.*, "Vehicular mobility simulation with VanetMobiSim", *Transactions of The Society for Modeling and Simulation International*, vol. 87, no. 4, 2011.

[HER 04] HERTKORN G., WAGNER P., "The application of microscopic activity based travel demand modelling in large scale simulations", *World Conference on Transport Research*, Istanbul, Turkey, 2004.

[ITE 11] ITETRIS, An integrated wireless and traffic platform for real-time road traffic management solutions, http://ict-itetris.eu/10-10-10-community/, 2011.

[JAA 05] JAAP S., BECHLER M., L.W., "Evaluation of routing protocols for vehicular ad hoc networks in city traffic scenarios", *IEEE ITSC*, Vienna, Austria, 2005.

[JET 03] JETCHEVA J., HU Y.-C., PALCHAUDHURI S., *et al.*, "Design and evaluation of a metropolitan area multitier wireless ad hoc network architecture", *IEEE WMCSA*, Monterey, CA, USA, 2003.

[JOE 12] JOERER S., DRESSLER F., SOMMER C., "Comparing apples and oranges? Trends in IVC simulations", *9th ACM International Workshop on Vehicular Internetworking (VANET 2012)*, Low Wood Bay, Lake District, UK, June 2012.

[KAR 07] KARNADI F., MO Z., LAN K.-C., "Rapid Generation of realistic mobility models for VANET", *IEEE WCNC*, Hong Kong, 2007.

[KRA 97] KRAUSS S., WAGNER P., GAWRON C., "Metastable states in a microscopic model of traffic flow", *Physical Review E*, vol. 55, no. 304, pp. 55–97, 1997.

[KRA 98] KRAUSS S., Microscopic modeling of traffic flow: Investigation of collision free vehicle dynamics, PhD thesis, University of Cologne, 1998.

[KRA 02] KRAJZEWICZ D., HERTKORN G., ROSSEL C., *et al.*, "SUMO (Simulation of Urban MObility): an open-source traffic simulation", *SCS MESM*, Sharjah, United Arab Emirates, 2002.

[LOC 03] LOCHERT C., HARTENSTEIN H., TIAN J., *et al.* "A routing strategy for vehicular ad hoc networks in city environments", *IEEE IV*, Columbus, OH, USA, 2003.

[MAT 12] MATSIM, http://matsim.org, 2012.

[MEI 11] MEIHUISOFTWARE, http://www.meihuichina.com, 2011.

[NAG 92] NAGEL K., SCHRECKENBERG M., "A cellular automaton model for freeway traffic", *Journal de Physique I*, vol. 2, no. 12, pp. 2221-2229, 1992.

[NAG 98] NAGEL K., WOLF D., WAGNER P., *et al.* "Two-lane traffic rules for cellular automata: a systematic approach", *Physical Review E*, vol. 58, pp. 1425–1437, 1998.

[NAN 06] NANDAN A., TEWARI S., DAS S., *et al.*, "AdTorrent: delivering location cognizant advertisements to car networks", *WONS*, Les Ménuires, France, 2006.

[OMN 12] OMNET++, http://www.omnetpp.org/, 2012.

[OPE 12] OPENSTREETMAP, The free wiki world map, http://www.openstreetmap.org, 2012.

[PAR 12] PARAMICS, http://www.paramics.com/, 2012.

[PIG 11] PIGNE Y., DANOY G., P. B., "A vehicular mobility model based on real traffic counting data", *Nets4Cars*, Oberpfaffenhofen, Germany, 2011.

[RAN 03] RANEY B., CETIN N., VÖLLMY A., *et al.*, "An agent-based microsimulation model of Swiss travel: first results", *Networks and Spatial Economics*, vol. 3, no. 1, pp. 23-41, 2003.

[RUD 02] RUDACK M., MEINCKE M., LOTT M., "On the dynamics of ad hoc networks for inter-vehicle communications (IVC)", *ICWN*, Las Vegas, NV, USA, 2002.

[SAH 04] SAHA A., JOHNSON D., "Modeling mobility for vehicular ad hoc networks", *ACM VANET*, Philadelphia, PA, USA, 2004.

[SES 92] SESKAR I., MARIE S., HOLTZMAN J., *et al.*, "Rate of location area updates in cellular systems", *IEEE VTC*, Denver, CO, USA, 1992.

[SOM 11] SOMMER C., GERMAN R., DRESSLER F., "Bidirectionally coupled network and road traffic simulation for improved IVC analysis", *IEEE Transactions on Mobile Computing*, vol. 10, no. 1, pp. 3-15, January 2011.

[SUM 12] SUMO, Simulator of urban mobility, http://sumo.sourceforge.net/, 2012.

[SWA 12] SWANS, Scalable wireless ad hoc network simulator, http://jist.ece.cornell.edu/, 2012.

[THA 11] THAKUR G., HUI P., KETABDAR H., *et al.*, "Towards realistic vehicular network modeling using planet-scale public webcams", *Computing Research Repository (CoRR)*, 2011.

[TIA 02] TIAN J., HAEHNER J., BECKER C., *et al.*, "Graph-based mobility model for mobile ad hoc network simulation", *SCS ANSS*, San Diego, CA, USA, 2002.

[TOM 10] TOMTOM, How TomTom's HD TrafficTM and IQ RoutesTM data provides the very best routing, White paper, 2010.

[TRA 12] TRANSIMS, Transportation analysis and simulation system, http://code.google.com/p/transims, 2012.

[TRE 00] TREIBER M., HENNECKE A., HELBING D., "Congested traffic states in empirical observations and microscopic simulations", *Physical Review E*, vol. 62, no. 2, pp. 1805–1824, 2000.

[TRE 02] TREIBER M., HELBING D., "Realistische mikrosimulation von strassenverkehr mit einem einfachen modell", *ASIM*, Rostock, Germany, 2002.

[TSI 12] TSIS-CORSIM, http://mctrans.ce.ufl.edu/featured/tsis/, 2012.

[UPP 11] UPPOOR S., FIORE M., "Large-scale urban vehicular mobility for networking research", *IEEE VNC*, Amsterdam, The Netherlands, 2011.

[VAN 12] VANETMOBISIM, The vehicular mobility simulator, http://sourceforge.net/projects/vanetmobisim/, 2012.

[VAR 06] VARSCHEN C., WAGNER P., "Mikroskopische Modellierung der Personenverkehrsnachfrage auf Basis von Zeitverwendungstagebüchern", *Stadt Region Land*, vol. 81, pp. 63–69, 2006.

[VEI 12] VEINS, http://veins.car2x.org/, 2012.

[VIR 11] VIRIYASITAVAT W., BAI F., TONGUZ O., "Dynamics of network connectivity in urban vehicular networks", *IEEE Journal on Selected Areas in Communications*, vol. 29, no. 3, pp. 515–533, 2011.

[VIS 12a] VISSIM, http://www.ptv-vision.com/en-uk/products/vision-traffic-suite/ptv-vissim/overview/, 2012.

[VIS 12b] VISUM, http://www.ptv-vision.com/en-uk/products/vision-traffic-suite/ptv-visum/overview/, 2012.

[VSI 12] VSIMRTI, http://www.dcaiti.tu-berlin.de/research/simulation/, 2012.

[WAG 99] WAGNER P., K.N., "Microscopic modeling of travel-deman: Approaching the home-to-work problem", *Transportation Research Board Annual Meeting*, Washington, DC, USA, 1999.

[WIE 74] WIEDEMANN R., "Simulation des straenverkehrsflusses", Schriftenreihe Heft 8, Institute for Transportation Science, University of Karlsruhe, 1974.

Chapter 7

Traffic Signal Control Systems and Car-to-Car Communications

7.1. Introduction

Since their inception in 1868 outside the British Houses of Parliament in London, traffic signaling systems have been one of the major components in traffic management strategies. In general, their purpose is to regulate the vehicular flow at the intersection by managing the movements of different vehicles approaching the intersection with reference to time. Precisely, the platoon of vehicles is allowed to cross the intersection in turn, or in phases, thus making it easier and safer for drivers as well as pedestrians. The basic concepts involved in any traffic signal systems are the following:

– Cycle: This is a complete sequence of all signal indications at an intersection.

– Phase: This is used to describe any part of a cycle allocated to any specific traffic movement at the intersection. A typical signal cycle with four phases is shown in Figure 7.1.

Chapter written by Mounir BOUSSEDJRA, Nitin MASLEKAR, Joseph MOUZNA and Houda LABIOD.

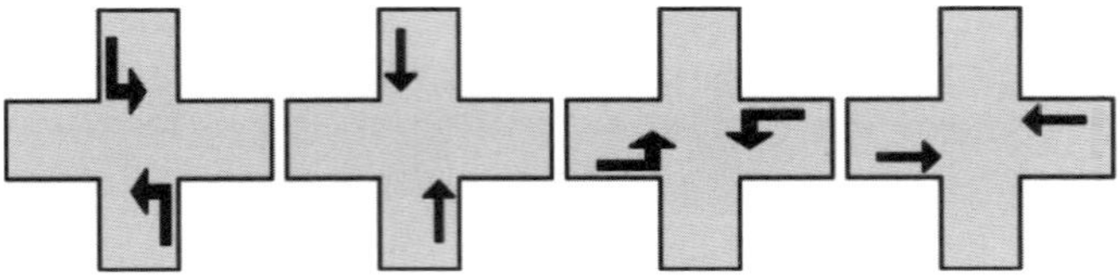

Figure 7.1. *Phases of signal cycle*

– Green split: This is a division of the cycle allocated to each of the various directions that will allow the passage of the vehicles.

– Yellow time: This is a safety interval by means of which road users are cautioned, because traffic lights are about to turn red or green.

– Red time: This is an interval during which signal indications at an intersection notify a stoppage of vehicles in one direction.

Most of the existing traffic signal systems work with a set of pre-timed cycles that generated off-line using the historical data of traffic demand. These systems try to match the best possible timing plan, from the existing database, for the current hour of the day. However, in recent years due to exponential increase of vehicular traffic, the traditional traffic signal control system fails to fulfill the requirements in urban areas for traffic management. Improper traffic signal control systems at the intersections are one of the major contributors for the increase in recurring traffic congestion [ABD 03]. Millions of people deal with such traffic congestion on a daily basis. Along with the effects on drivers, environment and health, such traffic congestion has shown its effect on economy as well. Within the US alone, the annual cost of congestion is estimated to be "US$ 63.1 billion", caused by "3.7 billion hours of delays" and "8.7 billion liters of 'wasted' fuel" [SCH 05], whereas in European Union the congestion costs nearly 1% of the EU's gross domestic product (GDP), annually [PRU 05]. Hence it is imperative to improve the efficiency of existing transportation facilities.

To ameliorate the traffic conditions in urban scenarios, it is important to develop dynamic traffic signal control that will adapt itself to varying vehicular traffic demand and optimize its timings accordingly. Such optimizations will result into smoother traffic flows at the intersections which will reduce the travel time for the users. This, in turn, reduces the number of vehicles that stop at the intersection and hence cuts the vehicular emissions.

To achieve this, many works have been proposed in the literature. In the following sections, a classification and description of the existing signal control methods are discussed.

7.2. Classification of traffic signal control systems

As discussed earlier, improper traffic signal controllers at the intersections are one of the factors that contribute to the degradation of the traffic flow. This drawback demonstrates the need for improvement in the functioning of traffic signal systems. Within this context, the technology of traffic signal system has evolved significantly over the years from the basic pre-timed systems to complex adaptive systems. The framework of the available solutions in traffic signal systems can broadly be classified into static systems and dynamic systems (Figure 7.2).

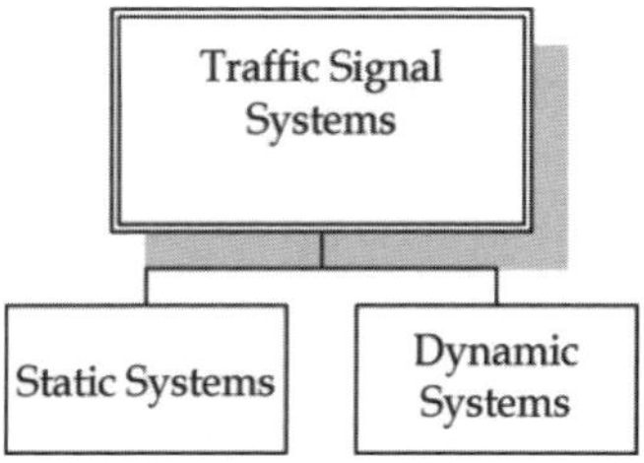

Figure 7.2. *Classification of traffic signal systems*

7.2.1. *Static systems*

Static systems represent the most basic form of traffic signal control (Figure 7.3) and operate on a predetermined regularly repeated sequence of signal indications [SKA 04]. The signal rotates through a defined cycle in a constant fashion, as determined by the controller's settings. For example, in one complete phase of the cycle, each street may be assigned some seconds of green time, several seconds of yellow interval and the remaining time to the red interval. This timing of static signals is typically determined from visual observations and traffic counts. Once the timing programs are set, they remain fixed until they are changed manually.

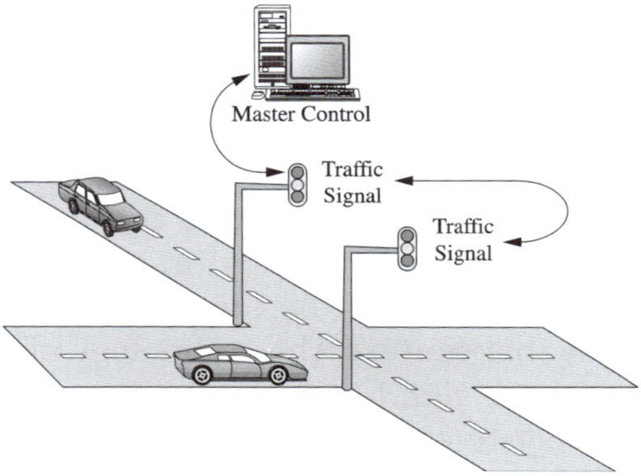

Figure 7.3. *Static signal system*

Static controllers are best suited for intersections where traffic volumes are predictable, stable and fairly constant. In general, static controllers are cheaper to purchase, install and maintain [GOR 96]. Their repetitive nature facilitates coordination with adjacent signals, and they are useful in areas where progression is desired. However, there is no guarantee that the system will have a plan for the existing traffic conditions. Thus, such systems fail to respond to the varying traffic demand and accommodate highly congested flow or non-recurring traffic conditions caused by accidents, weather conditions or special events [SKA 04].

7.2.2. *Dynamic systems*

The inflexibility of fixed systems prompts for the implementation of dynamic signal systems. These systems can accommodate not only recurring traffic congestion but also adequately adjust signal timing for non-recurring traffic congestion caused by random fluctuations in traffic patterns. Such systems are termed as dynamic system (Figure 7.4).

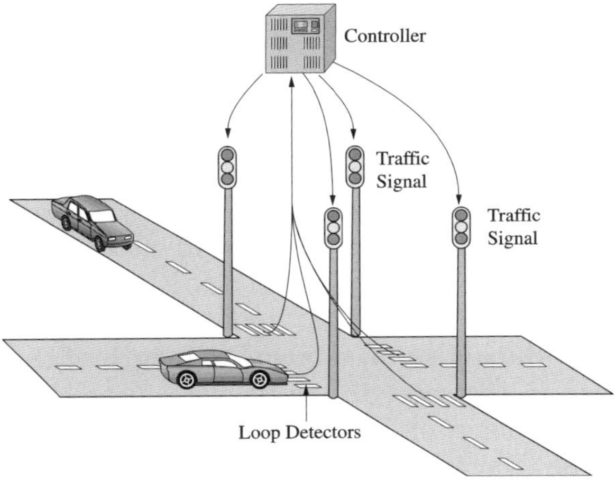

Figure 7.4. *Dynamic traffic signal systems*

These systems optimize the timing plans according to the vehicular demand and help ease the congestion and its negative externalities without the cost and environmental impact of road expansion. In the literature, there are several ways, such as inductive loops or sensor, through which the existing circumstances of the vehicular traffic can be procured [GOR 96]. On the basis of the working methodology, the dynamic systems can be further classified into three classes: actuated, responsive and adaptive

(Figure 7.5). Of these, adaptive systems are the most advanced and complex traffic signal systems.

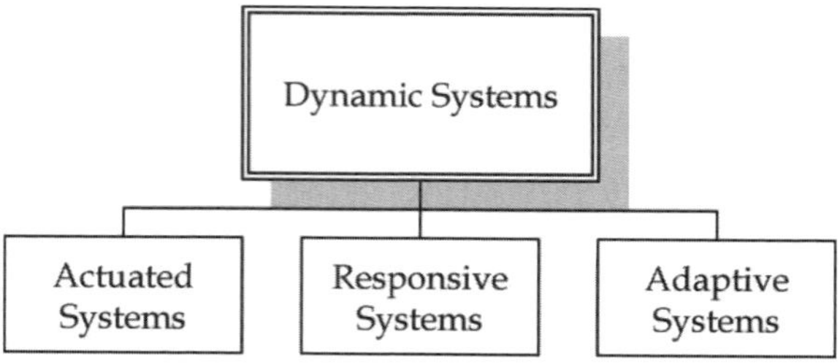

Figure 7.5. *Classification of dynamic systems*

7.2.2.1. *Actuated signal control*

Vehicle-actuated controllers are the basic dynamic systems that operate in real-time by applying a control in accordance to the current traffic state. An actuated controller operates based on traffic demands as registered by the actuation of vehicle and/or pedestrian detectors. The most common method of detecting vehicles is to install inductive loop wires [GOR 96]. The main feature of actuated controllers is the ability to adjust the length of the currently active phase in response to traffic flow. In such controllers, the green time for a phase which is a function of the traffic flow can be varied between pre-timed minimum and maximum lengths depending on flows. If there are no vehicles detected on an approach, the controller skips that phase. Actuated control can be further classified into semi-actuated control and full-actuated control.

A semi-actuated controller is usually installed in places where small streets join the main road. In this, the actuator can modify all phases except the phase associated with the main road. A continuous green time is maintained on the major street, that is the main road except when a demand is registered by the minor street or the joining street detector. Later the green time returns to the major street when no vehicles are present on the minor street or when the timing limit for the green period has been reached. A semi-actuated operation is best suited for locations with low volume of minor street traffic. It may also be used to permit pedestrian crossings mid street [BRU 94].

In full-actuated control, the function of the controller is to measure traffic flow on all approaches to an intersection and make assignments of green splits in accordance with traffic demand. Full-actuated control requires placement of detectors on all approaches to the intersection. The controller's ability to respond to traffic flow provides maximum efficiency at individual locations. This type of control is appropriate for intersections where the demand proportions from each leg of the intersection are less predictable.

Although vehicle-actuated controllers operate in real-time, they attempt no systematic optimization. Vehicle-actuated traffic light controllers are simply an enhanced version of fixed-time controller where the green light time is extended whenever there is a demand. These times are fixed prior to the maximum time limit and thus do not adapt well to varying traffic conditions. Hence there is a need for better solutions in dynamic traffic signal control.

7.2.2.2. *Responsive signal control*

Adhering to the principles of the dynamic systems, in responsive mode the traffic signal control receives information that reflects the current traffic conditions. On the basis of this input, the system selects an appropriate timing plan from the library of different plans, which is already present in the system. In responsive systems, usually decisions about cycle timings and the splits are based on some fuzzy logic rules [VAN 06]. Typically, in a fuzzy logic controller the extension time is not a fixed value. The number of cars sensed at the input of the fuzzy controllers is converted into fuzzy values, such as low, medium and high. These fuzzy values are then compared with a defined set of rules and then the splits are allotted to different phases. Figure 7.6 depicts a basic idea of fuzzy logic system. Here, the inputs which can be various factors like the volume of traffic are processed by the fuzzifier and are then matched to corresponding rules and are given to the defuzzifier. The defuzzification unit processes the data and produces a quantifiable result, in this case the signal timings.

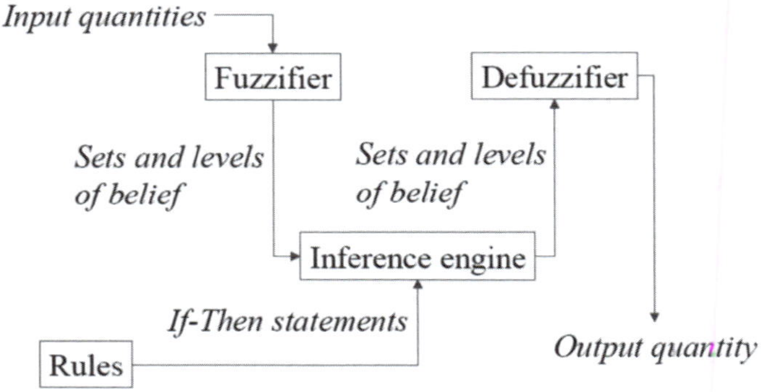

Figure 7.6. *Fuzzy logic system*

The authors [ANT 05] present a fuzzy logic based traffic signal systems where loop detectors are employed for estimating the number of vehicles on each lane. At the end of each phase these numbers are used as inputs to fuzzy controller that calculates the next signal cycle duration. At the fuzzy controller each input is represented with three membership functions corresponding to low, medium and high traffic density. Inputs processed through fuzzy controller rule to arrive at the green light duration for the forthcoming phase.

The analysis shows that real-time fuzzy logic controller for traffic lights signal durations in combination with optimized phase is an alternative to a classical fixed phase duration. However, this approach is tested only for an isolated intersection; hence it may not result into an effective green wave.

An enhancement to this approach is described in [FAH 06] where the author describes a fuzzy logic controller for a roundabout with four intersected roads. In this approach, there are two sensors placed on each incoming lane (Figure 7.7). The first sensor (downstream sensor) is at the traffic lights and counts the cars passing the traffic lights. The second sensor (upstream sensor), which is located before the first sensor, counts the cars coming to the roundabout at a given distance from the lights. The amount of cars for a roundabout lane is determined by the difference of the readings between the two sensors.

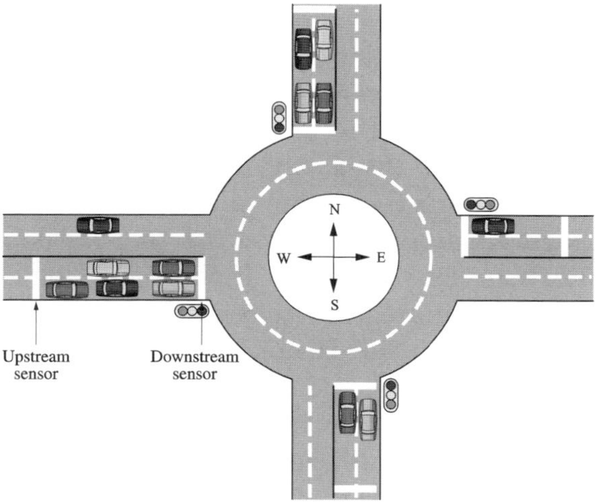

Figure 7.7. *Structure of fuzzy traffic control system*

Once the traffic volume is estimated the fuzzy logic controller determines which approach has the highest priority in order to give it the green light. This priority is decided by the fuzzy controller based on the following two factors: the quantity of the traffic waiting on the arrival side and the traffic waiting time.

The results show that the performance of this approach is better than the fixed-time controller, or even vehicle-actuated controllers. The reason is that it counts the number of vehicles sensed at the incoming approaches, and their waiting time then finds the winning approach for green light. Though this approach is better than the traditional systems, it has the drawback of starvation. There might be cases where only one lane gets the priority for all the cycles thus resulting in a degraded performance.

The authors [ZHA 05] propose a fuzzy logic controller to determine whether to extend or terminate the current green phase based on a set of fuzzy rules. The fuzzy rules compare traffic conditions with the current green phase and traffic conditions with the next candidate green phase.

The fuzzy logic controller determines whether to extend or terminate the current green phase after a minimum green time has expired. Once the green time is extended, after a specific time interval the fuzzy logic controller determines whether the current green time should be extended or not. This time interval may vary from 0.1 to 10 s depending on the controller's processor speed. If the fuzzy logic controller determines to terminate the current phase, then the signal will go to the next phase. If not, the current phase will be extended and the fuzzy logic controller will make the next decision after the next time interval and so forth until the maximum green time is reached. The decision-making process is based on a set of fuzzy rules that takes into account the traffic conditions with the current and next phases. Though the performance in terms of waiting time is improved as compared to the solution proposed by [FAH 06], the system is not totally dynamic as the green time cannot be extended beyond a certain maximum time and this extension limit is also fixed. Although this helps in avoiding starvation, such hard bound limits should be avoided to make the system more efficient.

This problem is tackled in [KHA 05] where the system considers two main features: first, to reduce the total delay time of waiting vehicles as well as to avoid heavy traffic congestion, and second to synchronize the local traffic controller with its neighbors, such as controlling the outgoing vehicles into neighboring traffic controllers. In this approach, three modules are proposed in the design of the fuzzy traffic lights controller: next phase module, a green phase module and a decision module (Figure 7.8).

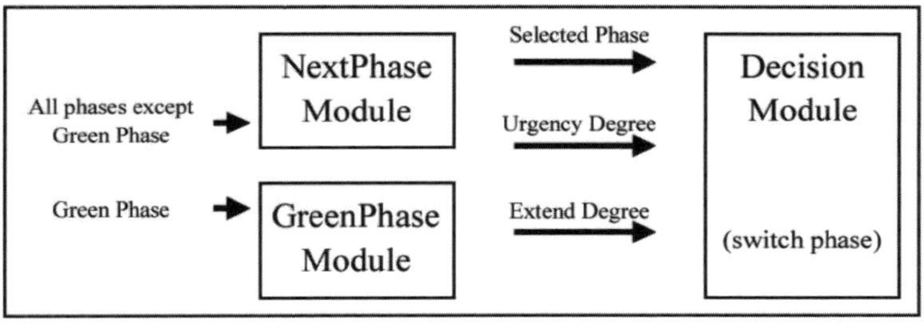

Figure 7.8. *Modules of the proposed fuzzy logic controller*

The next phase module and green phase module consider and evaluate the number of vehicles from the local detectors (detectors within the lane itself) and remote detectors (detectors from neighbor lanes). The next phase module selects the most urgent phase among all the other phases except the green phase. The green phase module observes the condition of traffic flow of the green phase only. The decision module decides the urgency degree between the next phase and the green phase modules. Thereby it can decide if the current green phase signal has to be extended or whether the passage should be given to other phases. For example, if the green phase module is more urgent than the next phase module, then the green signal will be extended. In addition, if the next phase module is more urgent than the green phase module, the decision module will change the green phase signal to another phase. As it is evident most of the systems are based on loop detectors and sensors.

However, the cost of deploying loop detectors is very high [DOA 09] thus making them an expensive approach.

An improvement to these systems is provided in [WEN 08] where an RFID-based detection technique for fuzzy logic approach is described. In this an RFID reader installed at the roadside detects an RF-ACTIVE code that is present on the vehicle. This allows us to detect the inter arrival rate of the vehicles on a given road. On the basis of these reading, a fuzzy controller selects the most appropriate rule that reflects the current condition and sets the green time for that particular phase. However in this work, the tests carried out are primitive and do not allow either two-way roads or car turning. The accuracy with which the RFIDs detect the flow of vehicles is also not tested thus making it vulnerable to degradation in efficiency.

To summarize the responsive systems it is seen that they use variables which allows for flexible implementation of the systems. The responsive systems allow for fuzzy thresholds which make them suitable in highly dynamic environment, because the rule based on which they work can be easily modified. However, the gradient of variation in the volume of vehicles can be very high, hence making it difficult to estimate membership function. In addition, there can be many ways of interpreting fuzzy rules, combining the outputs of several fuzzy rules and defuzzifying the output can be time consuming which can decrease the efficiency of the system.

7.2.2.3. *Adaptive signal control*

Adaptive traffic signal control systems have evolved into a potential technique for improving the traffic conditions in urban areas. They are currently the most advanced and complex dynamic control systems. They are similar to traffic responsive signals where they receive real-time data through some detectors. However, instead of matching current conditions to an existing timing plan, the system uses an optimization algorithm to create optimal signal timings in real-time. Hence, no libraries of timing plans are needed. This is advantageous for areas with high rates of growth, where libraries of timing plans have to be updated frequently. Many techniques have been developed to achieve adaptive traffic signal systems [LIU 07]. These existing adaptive systems can be divided into various approaches based on the mode they use to determine the timing plan (Figure 7.9).

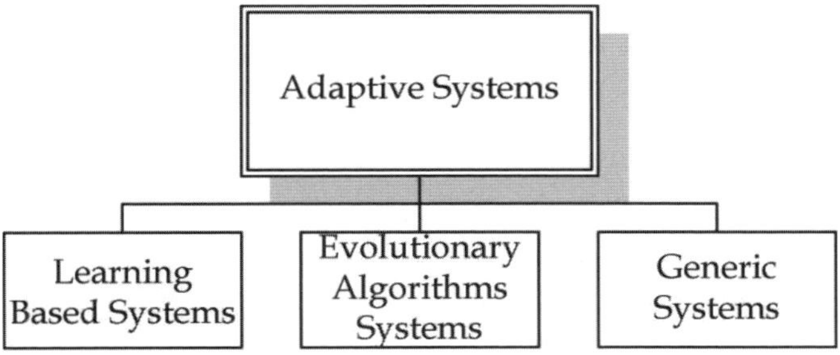

Figure 7.9. *Adaptive traffic signal systems*

7.2.2.3.1. Learning-based systems

In general, learning-based algorithms are concerned with how the system should take actions to maximize some notion of long-term reward. In this case of traffic signal systems, the long-term reward is in terms of reduction in waiting time for the road users. Figure 7.10 shows the basic concept of the learning system. In this representation, the agent is the entity responsible for interpreting sensory inputs from the environment, choosing actions and learning on the basis of the effects of its actions on the environment. On the basis of its perception of the state, the agent selects an action from the set of possible actions. This decision depends on the relative value of the various possible actions. As a result of taking action in state, the agent receives a reward that depends on the effect of this action on the agent's environment. The combination of state, action and reward is then used to update the previous estimate.

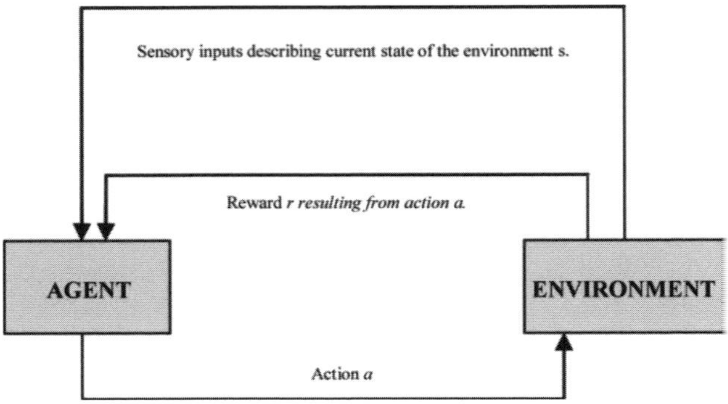

Figure 7.10. *Concept of learning-based systems*

One such adaptive control system based on learning algorithms is proposed by [KOO 04]. In this system, the authors develop a learning method with road-user based function to determine optimal cycles for each traffic light. This decision is based on a cumulative vote of all vehicles waiting at the intersection. For this, a value function is defined that estimates for each car the total waiting time until it has reached its destination. At the intersection the system computes the overall gain for each car if the lights were made green by traffic controller and their individual advantages are summed up.

In a similar way, the authors [GRE 07] describe an adaptive traffic light controller using intelligent agents at isolated intersections. In this system, the number of vehicle waiting near the intersection is detected through the sensors. On the basis of the number of vehicles, the idea here is to allow the agents (decision-maker at the traffic signal control) learn a traffic control policy using machine learning algorithms. This is aimed at reducing the waiting time for the vehicles at the intersection. To enable this, intelligent agents need a formal framework for decision-making. In this work, this is achieved through the use of the Markov decision processes (MDP) in the field of reinforcement learning [SUT 98]. To find the optimal solution of an MDP, the Markov property must be satisfied. This property indicates that the transition probability of reaching the next state of the traffic signal control only depends on the current state. With this property, the current state of the system encapsulates all the knowledge required in making a decision.

However, this algorithm induces high computational cost. To implement good timing patterns the agents will need to include many subtle details of the system and hence takes time to converge at the optimum timing pattern.

The authors [ELT 10] propose reinforcement-based adaptive traffic signal control with a variable phasing sequence. The authors define three evaluation functions: arrival of vehicles to the current green direction, queue length at red directions and cumulative delay. A variable phasing sequence is used to select the next action of the phase which will maximize the reward (defined as the decrease in the total cumulative delay). A greedy method [SUT 98] is used for action selection in which a greedy learner selects the action which will maximize the reward. The use of the greedy method enables the learning agent to search the overall state-action space and gradually converges to the optimal policy. The performance analysis showed that proposed approach consistently outperforms the pre-timed signal plan with a wide margin regardless of the demand level. However, since the greedy method is used to maximize the rewards, it might actually lead to starvation for a few phases approaching the intersection resulting in an inefficient system.

The authors [DUA 10] propose a new multi-objective control algorithm based on reinforcement learning for urban traffic signal control, named multi-RL. A multi-agent structure is used to describe the traffic system where vehicles are regarded as agents. A reinforcement learning algorithm is adopted to predict the overall value of the optimization objective for the given vehicles' states. The optimization objectives include the number of vehicle stops, the average waiting time and the maximum queue length at the intersection. To make the method suitable for various traffic conditions a multi-objective control scheme is introduced. Using the optimization objectives the system allows us avoiding large-scale traffic jams and hence a better traffic flow.

Learning algorithms are truly adaptive, in the sense that they are capable of responding not only to dynamic sensory inputs from the environment, but also to a dynamically changing environment, through ongoing learning and adaptation. In general, reinforcement learning does not require a pre-specified model of the environment on which action selection is based. Instead, relationships between states, actions and rewards are learned through dynamic interaction with the environment. Another benefit of reinforcement learning, in general, is that supervision of the learning process is not required. The outcome associated with taking a particular action in any state encountered is learned through dynamic exploration of alternative actions and observation of the relative outcomes.

However, exploration of the actions and the corresponding results are necessary in the initial stages of deployment of the system, and the learning process will be incremental in nature. This procedure of gaining useful experience while exploring actions may result in non-optimal solutions, which lead to high convergence time and computation cost for an optimal solution. Moreover, a sensory subsystem is required to provide the required inputs to the agents. This may involve adaptation of the existing induction loop technology, etc. Hence any loss of communication will result in a variation of the optimal policy. Therefore, a contingency plan would also be required to deal with a potential loss of communication.

7.2.2.3.2. Evolutionary algorithms

Evolutionary algorithms (EAs) are randomized optimization heuristics that mimic biological evolution to tackle optimization problems. Their general scheme is simple (Figure 7.11) which starts with a set (called population) of randomly generated initial solutions. An EA selects solutions with a relatively high quality from its population of parents, which are then combined and locally modified by crossover and mutation operators to form new offspring solutions. On the basis of their quality, some of the parents and offspring are selected to form the next generation of solutions that replaces the old population. This process is repeated until a stopping criterion is reached. Selection, crossover and mutation are randomized operations, where good solutions have a higher chance of surviving and generating offspring.

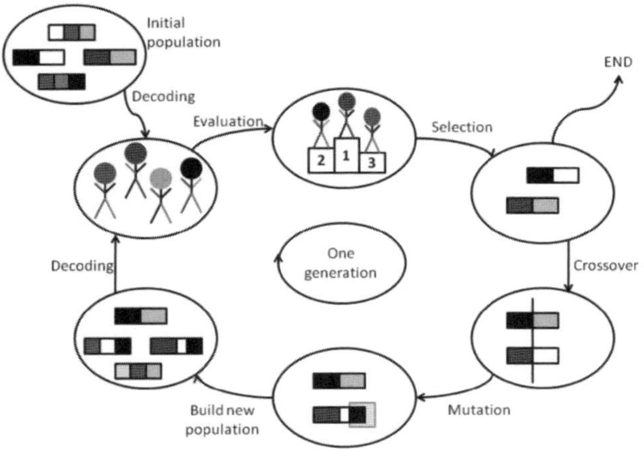

Figure 7.11. *General scheme of evolutionary algorithm*

EAs are widely used in many real-world optimization problems. Since the beginning of the 1990s they have also been applied to the optimization of traffic signal controls.

The research project TRAVOLUTION investigates the use of EAs for network-wide online optimizations [BRA 08]. The authors use an EA for the optimization of a signal plan that specifies the offsets, phase sequences and time frames bounding for possible phase endings. Optimization is aimed at minimizing a single-objective problem, that is the aggregate delay at the intersection. The proposed approach was evaluated in a field test at Ingolstadt, Germany. Using the EA, delays could be reduced by 21% compared to the static systems. This project is in its initial stages at the AUDI research and its being evolved into a more practical solution.

In [PRO 10], an EA is combined with a learning classifier system (LCS), a rule-based online learning mechanism, to form an adaptive learning intersection control system. While the LCS is responsible for the online selection of signal plans (based on detected traffic flows at the intersection), an EA is used to evolve necessary signal plans that are included in the LCS rule set. The study showed that delays could be reduced by 6–12%.

The simple working principle of EAs can be applied to any problem where a quality can be assigned to a solution. Therefore, the overall quality of solutions is likely to improve over time while the random influence of mutation helps to prevent premature convergence on some local optimum. Though this is an advantage, it may lead to high convergence time to an optimal solution that can hamper the efficiency in fast varying conditions.

7.2.2.3.3. Generic methods

Learning-based algorithm and EA form a broader category of methods for adaptive traffic signals control. However, there are other approaches as well which provide adaptive solutions, but they do not classify into either of the two. These solutions range from queuing theory-based to wireless system-based applications.

In [KUT 06], an adaptive algorithm is defined based on queuing theory for traffic signal control. Here an extended queue is modeled as a vector denoting the vehicles that have been waiting at the intersection. Including the mean waiting times in the model allows for a fair traffic control. The goal of the control is to find an optimal schedule for the traffic lights in the intersection, such that the difference in average waiting times between the two queues is minimized. For this, the authors apply a nonlinear model predictive controller (NMPC) [HEN 98]. NMPC allows for modifying the controller by taking into account the future incoming flow. This allows for the controller to prepare for a much better control. In addition, the use of queuing model and the use of waiting time in the state reduce the computation costs, however, it still takes time to evolve to an optimal solution.

Another mechanism of adaptive traffic signal based on constraint satisfaction (CSP) is mentioned in [MIZ 08]. Here the signal parameters, the cycle length, the split and the offset are defined as the CSP, and the optimal cycle length is determined by solving CSP with reduction in waiting time as the constraint. Precisely, the constraints are defined as:

$$P = (V, D, C)$$

where V is a set of variables, in which elements: cycleik, gstartik and gtimeik, correspond to time length of one cycle, start time period of the green sign in the cycle and time length of the green sign, respectively, in the kth signal belonging to the ith intersection agent; D is a set of values to be assigned to variables, where Dcycle = {40, ..., 140} and Dgstart = Dgtime = {0, ..., 140}. Therefore, a signal has cycle time length from 40 s to 140 s and green time length from 0 s to 140 s. C is a set of constraints composed of subsets, Cin, Cadj and Croad.

Though the simulation results show that the traffic jam is considerably reduced, but solving the CSP incurs time to arrive at an optimal solution. In addition, the length of the cycle time is bounded by preset values, hence in case of severe variation the system may fail to adapt.

To reduce the convergence time of obtaining an optimal solution, the authors [ALI 06] define a greedy control with scarcity measurement approach (GSCM). In GSCM, the control method is similar to many feedback algorithms. This feedback is quantized to a value of resource scarcity. The resource here is the streets which are to be assigned to the road users. The decision taken to set the lights green is based on the following formula:

$$\begin{aligned}&\text{StateChange} = \text{Min}\{|\,\Delta\text{price open street}\,|\} - \text{Max}\begin{Bmatrix}\text{Max}\{|\,\Delta\text{price open street}\,|\},\\ \text{Max}\{|\,\Delta\text{price downstream streets}\,|\}\end{Bmatrix}\\&\text{StateChange} \le 0 \rightarrow \text{Better Changes}\\&\text{StateChange} > 0 \rightarrow \text{Worsening Change}\end{aligned} \quad [7.1]$$

Here the authors explain the price of a resource as the state of scarcity of resource and is given by:

$$price = 1000 * \log \frac{|\,ls + \varepsilon}{nss + \varepsilon} \cdots \quad [7.2]$$

where *ls* is the load on the street and nss is the normalized mean speed of the street and ε is set between 0 and 1.

GCSM control algorithm prohibits the control agents making worsening changes which allow us handling the traffic at the intersection in a much simpler way. However, the function has imposed two restrictions: (1) the street light changes to green in a predefined order and (2) the street green light length must be in interval of "minimum green time, maximum green time". Such restrictions nullify the effectiveness of the system. In addition, it adapts greedy approach to select the changes that may lead to too many fluctuations if vehicular traffic undulates severely.

The rigidity and aforementioned drawback of the traditional adaptive systems prompts the exploration of a new domain to realize adaptive traffic signal controls. With recent advances in communication, computation and sensor technologies a new era of intelligent transportation using wireless communication can be envisioned. A protocol based on wireless technology is introduced in [HUA 04]. In this, the authors propose a concept of smart intersection using wireless communication. Each car sends its information, such as speed and distance to the intersection, to the traffic signal. Using this information, traffic signal control sends the decision if a car can pass the intersection or not. On the basis of this response, the car adjusts its speed so that it can reduce the waiting time at the intersection. This method aids arrival at optimal decisions in a decentralized manner, because each vehicle decides on its approaching speed to the intersection. It is thus facilitated for a faster convergence but it is prone to failure of data communication which may cause the whole system to collapse.

Although the domain of wireless communications is in its primitive stages, the work mentioned by [HUA 04] provides the impetus to investigate the domain of wireless technologies for adaptive traffic control systems. Accordingly, the authors [ZHO 10] investigate the problem of adaptive traffic light control using real-time traffic information collected by a wireless sensor network (WSN). The authors propose an adaptive traffic light control algorithm that adjusts both the sequence and length of traffic lights in accordance with the real-time traffic detected. The algorithm considers a number of traffic factors, such as traffic volume, waiting time, to determine green light sequence and the optimal green light length. The green light length is based on the different weights assigned to traffic volume, average waiting time, hunger level, blank circumstance, special circumstance and influence from neighboring intersections. The use of blank circumstances allows for fine tuning the green time so that when a blank situation approaches the intersection it is dealt with a red timing. However, the green time distribution is also based on hunger level, which indicates how much traffic is on an approaching lane. This factor can influence starvation because the less dense roads may be totally deprived of green time.

Along with sensor networks, vehicular networks also form a capable solution. The authors [GRA 07] defined one proposed solution where one-hop car-to-car communications are used for implementing the traffic controls. The authors use the solution implemented in TrafficView [NAD 04], wherein the vehicles exchange the information about the position and the speed among themselves. This enables each individual vehicle to view and assess the traffic and road condition in front of it. To achieve data dissemination the diffusion mechanism is utilized where each vehicle broadcasts information about itself and the other vehicles it knows about. Whenever a vehicle receives broadcast information, it updates its stored information and defers forwarding the information to the next broadcast period, at which time it broadcasts its updated information. During this exchange, the traffic signals listen to the communication between cars and estimate the density of vehicles around it and adjust the signal timings accordingly. The results show that the system significantly improves traffic fluency, compared to the existing pre-timed traffic lights.

7.3. Traffic signal control and car-to-car communication

As discussed in previous sections, to efficiently implement adaptive traffic control systems, we must take into account a wide range of variables, such as traffic volume, time of the day, the effects on other roads and the involvement of pedestrian traffic. In this context, vehicular *ad hoc* networks (VANETs) can be quite effective in estimating the volume of the traffic approaching the intersection. One of the methods called the car–car based adaptive traffic control system (CATS) is discussed in [MAS 12]. To accomplish this through VANETs, it is essential to have an effective data dissemination strategy. In the proposed approach, a cluster-based data dissemination protocol called clustering based on direction in vehicular environment (C-DRIVE) [MAS 11]. Performance evaluation of the system shows that the average waiting time is considerably reduced compared to the fuzzy logic and pre-timed systems. In terms of efficiency, as shown in Figure 7.11, it is found that CATS reduces the number of vehicles that stop at the intersection by almost 20%. The encouraging results show that adaptive traffic lights based on car-to-car communication has a good potential for improving the traffic conditions in urban areas.

7.4. Summary and conclusion

Traffic signal systems are an important element to improve the traffic flow in urban scenarios. Their basic aim is to regulate the traffic at intersections. Along with facilitating a smooth traffic flow that will save a lot of time and fuel for road users, traffic lights also help in reducing the number of accidents. Traditional traffic signal systems work in a pre-timed manner in which the cycle timings are preset. Such rigidity does not work well in the fast-growing urban scenarios. The need of the hour, in fast varying traffic conditions, is to provide systems that can change their cycle timings dynamically.

With this objective much work has been done in this area where signal timings dynamically adjust to the traffic demand. The various dynamic systems can be classified into three major categories: actuated, responsive and adaptive systems. Adaptive are the most complex and advanced systems. They can be further classified into learning based, EAs and generic methods. Most of the works mentioned in generic methods are based on the use of wireless communications. As compared to the traditional pre-timed systems the dynamic systems help in improving the traffic conditions at the intersections. Precisely, dynamic systems improve the responsiveness of signal timing in rapidly changing traffic conditions. Various works have demonstrated network performance enhancement from 5% to over 30% in the user waiting time at the intersection.

Though various methods defined under dynamic systems show improvement in the traffic efficiency, they still leave a lot of scope for development. Table 7.1 gives a brief summary of the classification along their advantages and disadvantages.

Apart from the implementation issues two factors, namely average waiting time and reduction in queue length form the basis on which the performance of the dynamic system is evaluated. Here, queue length represents the number of vehicles that stops at the intersection when they encounter a red light. The average waiting time factor presents the delay experienced by the vehicles that stops at the intersection. Table 7.2 provides a comparative analysis for various dynamic systems in terms of these two factors.

Signal systems	Subclass	Property	Advantages	Disadvantages
Pre-timed		Static – cycles are preset in the system	Easy to implement	Cannot adapt to the varying traffic demands
Actuated	Semi-actuated	It is user controlled and usually implemented when the lanes connecting to the main road are of not much importance	User actuated so no central control unit required	Cannot be implemented on a large scale for adaptive signals
	Full-actuated	Uses techniques like loop detectors to measure traffic flow and make assignments in accordance with traffic demand	Can adapt easily to the varying traffic demands at the intersections	Implementation costs of techniques such as loop detectors is high
Responsive	Fuzzy logic	Decisions are made based on the qualitative knowledge about the traffic	Easy algorithms to implement the adaptive behavior of traffic signals	Based on too much rule so can fail in adverse situations
Adaptive	Learning-based	In general, the learning-based systems work on the principle to maximize gain desired for the result	Policy-driven approach so easy to adapt to varying traffic demands	High computational costs
	Evolutionary algorithms	Such algorithms evolve over a period of time to attain the best possible result	Work well in scenarios where the traffic patterns show a gradual change	High convergence time Cannot adapt rapidly to the dynamic environment

Table 7.1. *Summary of traffic signal control systems*

Dynamic systems	Effect on average waiting time and queue length
Actuated	Since they are extension of pre-timed systems, waiting time can be reduced to an extent only for recurring traffic congestion. This is achieved by extending green time to a certain limit. The queue length can be controlled for recurring traffic congestions, but it tends to increase exponentially for non-recurring and volatile traffic conditions.
Responsive	In this mode, the system selects an appropriate timing plan from the library of different plans which are already present in the system. Hence, the waiting time and the queue length will be reduced only if exact criteria reflecting the traffic conditions are found.
Learning-based	In learning-based systems, the waiting time can be reduced only after the systems stabilize to an optimum output. However, to arrive at stability point it takes time and hence the average gain in terms of waiting time and reduction in queue length is limited during the initial phase. Even after the stable state, if the traffic conditions change the system reinitiates the learning process thus incurring more computational costs.
Evolutionary	The performance of evolutionary algorithm is similar to learning-based systems. They achieve better results only if the changes are gradual and not abrupt.
Wireless systems	These are new generation systems in which the data about the vehicular traffic are aggregated through wireless communications and then the signal timings are calculated according to the received data. This approach helps to adapt to recurring as well as non-recurring traffic and thus helps in improving the waiting time and the queue length

Table 7.2. *Comparison of dynamic systems*

It is pretty much evident that dynamic traffic signal controls are capable of addressing the demands of varying traffic conditions. However, there is still a lot of scope for improvement. This can be achieved by exploring various possible avenues. Among all, wireless VANETs are emerging as an effective solution for intelligent transportation systems of which traffic signal control is a part. VANETs can provide real-time information of the volatile traffic and are thus perceived to enable signal control systems to operate with greater efficiency. In addition, sharing traffic signal and operations data with other systems will improve overall transportation system performance. Few proposed works discussed in this literature [MAS 12], which was discussed in section 7.3, show how the use of vehicular networks can improve the traffic flow at the intersection. However, as with any other technique to implement adaptive traffic lights using VANETs, the issues pertaining to it have to be addressed.

7.5. Bibliography

[AAS 01] AASHTO, Standard Specification for Structural Supports for Highway Sign, Luminaries and Traffic Signals, American Association of State Highway and Transportation Officials, 2001.

[ABD 03] ABDULHAI B., PRINGLE R., KARAKOULAS G.J., "Reinforcement learning for true adaptive traffic signal control", *Journal of Transportation Engineering*, vol. 129, pp. 278, 2003.

[ALI 06] ALIPOUR M.A., JALILI S., "Urban signal control using intelligent agents", *Applied Artificial Intelligence 7th International FLINS Conference*, World Scientific Publishing Company, pp. 811–816, 2006.

[ANT 05] ANTUNOVIĆ M., GLAVAŠ H., "Fuzzy logic approach for traffic signals control of an isolated intersection", *Skup o prometnim sustavima s međunarodnim sudjelovanjem* AUTOMATIZACIJA U PROMETU 2005.

[BRA 08] BRAUN R., KEMPER C., WEICHENMEIER F., "TRAVOLUTION: Adaptive urban traffic signal control with an evolutionary algorithm", *Proceedings of the 4th International Symposium "Networks for Mobility"*, Stuttgart, September 25th and 26, 2008.

[BRU 94] BRUNO G., IMPROTA G., BIELLI M., *et al.*, "Urban traffic control: current methodologies", *Artificial Intelligence Applications to Traffic Engineering*, pp. 69–93, 1994.

[DOA 09] DOAN Q.C., BERRADIA T., MOUZNA J., "Vehicle speed and volume measurement using vehicle-to-infrastructure communication", *WSEAS Transactions on Information Science and Applications*, vol. 6, no. 9, pp. 1468–1477, 2009.

[DUA 10] DUAN H., LI Z., ZHANG Y., "Multiobjective reinforcement learning for traffic signal control using vehicular Ad Hoc network", *EURASIP Journal on Advances in Signal Processing*, vol. 2010, pp 7–17, 2010.

[ELT 10] El-TANTAWY S., ABDULHAI B., "An agent-based learning towards decentralized and coordinated traffic signal control", *Intelligent Transportation Systems (ITSC), 2010 13th International IEEE Conference on*, IEEE, pp. 665–670, 2010.

[FAH 06] FAHMY M.M.M., "An adaptive traffic signaling for roundabout with four approach intersections based on fuzzy logic", *Journal of Computing and Information Technology*, vol. 15, no. 1, pp. 33–45, 2006.

[GOR 96] GORDON R.L., TIGHE W., UNITED STATES FEDERAL HIGHWAY ADMINISTRATION. OFFICE OF OPERATIONS, UNITED STATES FEDERAL HIGHWAY ADMINISTRATION, OFFICE OF TRANSPORTATION MANAGEMENT, DUNN ENGINEERING ASSOCIATES, AND ITS SIEMENS, Traffic control systems handbook, Department of Transportation, Federal Highway Administration, Office of Technology Applications, 1996.

[GRA 07] GRADINESCU V., GORGORIN C., DIACONESCU R., *et al.*, "Adaptive traffic lights using car-to-car communication", *IEEE 65th Vehicular Technology Conference 2007 (VTC2007-Spring)*, IEEE, pp. 21–25, 2007.

[GRE 07] GREGOIRE P.L., DESJARDINS C., LAUMÔNIER J., *et al.*, "Urban traffic control based on learning agents", *IEEE Intelligent Transportation Systems Conference, 2007 (ITSC 07)*, IEEE, pp. 916–921, 2007.

[HEN 98] HENSON M.A., "Nonlinear model predictive control: current status and future directions", *Computers and Chemical Engineering*, vol. 23, no. 2, pp. 187–202, 1998.

[HOS 10] HOSSEIN P.N., SHAHROKH V., MAZIAR N., "Vehicular ad hoc networks", *EURASIP Journal on Advances in Signal Processing*, vol. 2010, pp. 1–2, 2010.

[HOU 01] HOU T.C., TSAI T.J., "A access-based clustering protocol for multihop wireless ad hoc networks", *IEEE Journal on Selected Areas in Communications*, vol. 19, no. 7, pp. 1201–1210, 2001.

[HUA 04] HUANG Q., MILLER R., "Reliable wireless traffic signal protocols for smart intersections", *ITS America Annual Meeting*, San Antonio, Texas, 2–4 May, 2004.

[JER 07] JERBI M., SENOUCI S.M., RASHEED T., *et al.*, "An infrastructure-free traffic information system for vehicular networks", *IEEE 66th Vehicular Technology Conference 2007 (VTC-07)*, IEEE, pp. 2086–2090, 2007.

[KHA 05] KHALID M., LIANG S.C., YUSOF R., "Control of a complex traffic junction using fuzzy inference", *Control Conference 2004 5th Asian*, vol. 3, IEEE, pp. 1544–1551, 2005.

[KOO 04] KOOPMAN A., VREEKEN J., VEENEN J., *et al.*, Intelligent traffic light control, Technical report UU-CS-2004-029, Utrecht University, 2004.

[KUT 06] KUTIL M., HANZÁLEK Z., CERVIN A., "Balancing the waiting times in a simple traffic intersection model", *Proceedings of 11th IFAC Symposium on Control in Transportation Systems 2006*, Citeseer, Netherlands, 29–31 August 2006.

[LEV 98] LEVINSON D.M., "Speed and delay on signalized arterials", *Journal of Transportation Engineering*, vol. 124, no. 3, pp. 258–263, 1998.

[LIU 07] LIU Z., "A survey of intelligence methods in urban traffic signal control", *IJCSNS*, vol. 7, no. 7, pp. 105–112, 2007.

[MAS 11] MASLEKAR N., BOUSSEDJRA M., MOUZNA J., *et al.*, "C-DRIVE: clustering based on direction in vehicular environment", *Proceedings of IEEE International Conference of New Technologies, Mobility and Security (NTMS)*, Paris, 7–10 February 2011.

[MAS 12] MASLEKAR N., BOUSSEDJRA M., MOUZNA J., et al., "CATS: an adaptive traffic signal system based on car-to-car communication", *Journal of Network and Computer Applications*, 2012.

[MIZ 08] MIZUNO K., FUKUI Y., NISHIHARA S., "Urban traffic signal control based on distributed constraint satisfaction", *Proceedings of the 41st Annual Hawaii International Conference on System Sciences*, IEEE, pp. 65–70, 2008.

[NAD 04] NADEEM T., DASHTINEZHAD S., LIAO C., *et al.*, "TrafficView: traffic data dissemination using car-to-car communication", *ACM SIGMOBILE Mobile Computing and Communications Review*, vol. 8, no. 3, pp. 6–19, 2004.

[PRO 10] PROTHMANN H., ROCHNER F., TOMFORDE S., *et al.*, "Organic control of traffic lights", *Autonomic and Trusted Computing*, Springer, vol. 1, pp. 219–233, 2010.

[PRU 05] PRUD'HOMME R., "The current EU transport policy in perspective", *Conference on European Transport Policy in the European Parliament*, European Commission, Brussels, p. 3, 2005.

[SCH 05] SCHRANK D.L., LOMAX T.J., TEXAS TRANSPORTATION INSTITUTE, *The 2005 Urban Mobility Report*, Texas Transportation Institute, Texas A & M University, 2005.

[SKA 04] SKABARDONIS A., "Traffic signal control systems", *Assessing the Benefits and Costs of ITS*, Springer, vol. 1, pp. 131–144, 2004.

[SUT 98] SUTTON R.S., BARTO A.G., *Reinforcement Learning: An Introduction*, The MIT press, 1998.

[VAN 06] VAN KATWIJK R.T., DE SCHUTTER B., HELLENDOORN J., "Traffic adaptive control of a single intersection: a taxonomy of approaches", *Proceedings of the 11th IFAC Symposium on Control in Transportation Systems*, Elsevier, IFAC Papers Online, pp. 227–232, 2006.

[WEN 08] WEN W., "A dynamic and automatic traffic light control expert system for solving the road congestion problem", *Expert Systems with Applications*, vol. 34, no. 4, pp. 2370–2381, 2008.

[ZHA 05] ZHANG L., LI H., PREVEDOUROS P.D., "Signal control for oversaturated intersections using fuzzy logic", *TRB 2005 Annual Meeting CD-ROM*, Washington DC, 9–13 January, 2005.

[ZHO 10] ZHOU B., CAO J., ZENG X., *et al.*, "Adaptive traffic light control in wireless sensor network-based intelligent transportation system", *Vehicular Technology Conference Fall (VTC 2010-Fall), 2010 IEEE 72nd*, IEEE, pp. 1–5, 2010.

List of Authors

André-Luc BEYLOT
ENSEEIHT, IRIT
Toulouse
France

Mounir BOUSSEDJRA
ESIGELEC
Rennes
France

Anthony BUSSON
IUT–University of Lyon 1
Laboratoire de l'Informatique et du Parallélisme (LIP)
École Normale Supérieure
Lyon
France

Emmanuel CHAPUT
ENSEEIHT, IRIT
Toulouse
France

Xunxing DIAO
Laboratoire LIMOS
ISIMA
Blaise Pascal University
Clermont-Ferrand II
France

Marco FIORE
CITI/UrbaNet
INSA Lyon/INRIA
Lyon
France

Yvon GOURHANT
Orange Labs
Lannion
France

Jérôme HÄRRI
Eurécom
Sophia-Antipolis
France

François JAN
Orange Labs
Lannion
France

Colette JOHNEN
LaBRI
University of Bordeaux 1
France

Florent KAISSER
IFSTTAR
Villeneuve d'Ascq
France

Houda LABIOD
Telecom ParisTech
Département INFRES
Paris
France

Jian-Jin LI
Laboratoire LIMOS
ISIMA
Blaise Pascal University
Clermont-Ferrand II
France

Nitin MASLEKAR
NecLab
Germany

Kun-Mean MOU
Laboratoire LIMOS
ISIMA
Blaise Pascal University
Clermont-Ferrand II
France

Joseph MOUZNA
ESIGELEC
Rennes
France

Guillaume RÉMY
Orange Labs
Lannion
France

Sidi-Mohammed SENOUCI
Laboratoire DRIVE
University of Burgundy
Nevers
France

Razvan STANICA
INSA Lyon
France

Nabil TABBANE
MEDIATRON
Sup'Com
Tunis
Tunisia

Sami TABBANE
MEDIATRON
Sup'Com
Tunis
Tunisia

Sandesh UPPOOR
CITI/UrbaNet
INSA Lyon/INRIA
France

Véronique VÈQUE
University of Paris-Sud
Orsay
France

Ghayet el mouna ZHIOUA
MEDIATRON/InfRes
Sup'Com/Telecom ParisTech
Tunis/Paris
Tunisia/France

Haiying ZHOU
School of Computer Science &
Technology
Harbin Institute of Technology
China

Index

K

L

M

N

O

P

Q

R